RECLAIMING AMERICAN CITIES

RECLAIMING AMERICAN CITIES

The Struggle for People, Place, and Nature since 1900

Rutherford H. Platt

University of Massachusetts Press • Amherst and Boston

Publication of this book has been assisted by a generous grant from
Yuji and Lorraine Suzuki

ISBN 978-1-62534-050-4 (paper); 049-8 (hardcover)

Set in Minion Pro and ITC Franklin Gothic
Printed and bound by IBT/Hamilton, Inc.

Library of Congress Cataloging-in-Publication Data

Platt, Rutherford H.
  Reclaiming American cities : the struggle for people, place, and nature since 1900 /
Rutherford H. Platt.
    pages   cm
  Includes bibliographical references and index.
  ISBN 978-1-62534-050-4 (pbk. : alk. paper)—ISBN 978-1-62534-049-8 (hardcover : alk. paper)
1. Urbanization—United States. 2. Urban policy—United States. 3. City planning—United States.
4. Urban ecology (Sociology)—United States. I. Title.
  HT384.U5P55 2014
  307.760973—dc23
   2013031756

British Library Cataloguing-in-Publication Data
A catalogue record for this book is available from the British Library.

To Barbara, always

Contents

Preface

This book grows out of the earlier book and series of conferences based on it titled *The Humane Metropolis: People and Nature in the 21st Century City*. The twenty-seven contributors to that volume collectively introduced a new way of looking at cities, through the prism of people, place, and nature as the elements of more "humane" urban communities and regions. This perspective, which will be called "humane urbanism" in this book, refers to grassroots efforts, often supported by government, nongovernmental organizations, and foundations, to make urban places at various scales greener, safer, more healthy, efficient, equitable, and people-friendly.

This book argues that the 1990s marked a turning point or "sea change" from nearly a century of top-down, expert-driven, "one-size-fits-all" urban policies towards more protean, bottom-up sets of initiatives better adapted to local needs and resources. The old top-down order (which of course persists in various forms today) focused on the downtown, mega-projects, the automobile, and wealth generation for the white establishment. By contrast, this new era is more concerned with uplifting the lives of ordinary people, reinvigorating the places where they live, and rehabilitating local parks, streams, waterfronts, forests, and other natural phenomena in their midst.

A bit of personal history to explain where this book comes from. The whole thing may have actually begun in utero and during my early childhood when my father (also Rutherford Platt but without the middle initial "H.") was avidly becoming a self-taught naturalist based, oddly at the time, in New York City. As I grew up, he was constantly rushing off lugging cameras, specimen cases, and notebooks to botanize in New England, the American West, Greenland, New Zealand, Australia, and other far-flung settings. His writings and photographs appeared in *Life, National Geographic, Scientific American, Readers Digest,* and his own many books during the 1940s into the 1970s. Meanwhile, back home in Manhattan, I joined my schoolmates in prowling the streets, courtyards, and rooftops of our neighborhood and riding the subway to Yankee Stadium, Central Park, and Coney Island. I was thus a young oxymoron, pulled in one direction by my father's fascination with Nature in the raw and in another by the city around me, its crowds, its parks, its sheer vitality.

After college and a couple of years on a U.S. Navy icebreaker (I was also drawn to the polar regions), I entered the University of Chicago Law School where I became a geographer. This unexpected outcome (rather upsetting to my parents) was inspired by the posthumous influence of my uncle, Robert S. Platt, a beloved professor of geography at the university, and of his indefatigable wife, Harriet S. Platt, an ardent champion of doing your own thing. Finding torts, contracts, and wills to be not my thing, I gravitated across campus to fraternize with the geography faculty and graduate students. I finished law school, taking everything available on local government, property, and the environment before crossing the Midway to pursue a doctorate in geography with a fellowship from the new U.S. Department of Housing and Urban Development. The geography program introduced me to Gilbert F. White, a product of the university's progressive tradition and exemplar of the scientist in service to government and public policy. With many others, I regarded Gilbert as a role model, a demanding mentor, and later a dear friend until his death in 2006.[1]

Further enriching the mix, during my final year in law school I met Gunnar A. Peterson, a genial Quaker who directed a fledgling organization called the Chicago Open Lands Project (later renamed Openlands). Gunnar hired me to help with legal research, first on a part-time basis and later, after I passed the bar exam, as full-time staff attorney. It proved to be a fruitful opportunity: my geography and law interests found a happy synthesis in my Openlands work, which in turn inspired and informed my doctoral thesis, *The Open Space Decision Process*. My main job at Openlands was to help local conservationists save threatened patches of prairie, wetlands, dunes, and forests scattered around metropolitan Chicago (such as Thorn Creek Woods, which I discuss in chapter 5). This opportunity coincided with comparable efforts in cities and suburbs across the country stimulated by William H. Whyte's famous 1968 book, *The Last Landscape*.[2] I also tried to be a nuisance regarding some major proposed regional projects, such as the city's scheme for an airport in Lake Michigan (that's correct) and a Crosstown Expressway through black neighborhoods on the city's south and west sides. Neither was ever built. (Part of my memo on the Crosstown appears in chapter 3).

Working in downtown Chicago in the late 1960s was exhilarating: civil rights demonstrations, the *Gautreaux* litigation (see chapter 3), the 1968 Democratic Convention protests, and the "Chicago Seven" trial were all at our doorstep. Chicago was an epicenter of movements rattling the nation: civil rights, antiwar, feminism, environmentalism, gay rights. This book in part reflects my own journey from a conventional Republican upbringing (as a kid I liked Ike, but who didn't?) to a fervent liberal, inherently skeptical of authority, especially when it is mindless or harmful.

In the 1980s, while pursuing a respectable career as a professor at the University of Massachusetts Amherst, I became a bit nostalgic for my Chicago adventures and friends. With support from the U.S. Forest Service Research Program, our program organized the Sustainable Cities Symposium held at the Chicago Academy of Sciences in October 1990 with William H. ("Holly") Whyte as keynote speaker. That conference eventually yielded *The Ecological City: Preserving and Restoring Urban Biodiversity* (1994)—the first University of Massachusetts Press prequel to this book. With a fertile mix of authors and topics, the book proved to be a hit in the new world of urban ecology.

I continued this line of effort with a new program of research, teaching, and outreach called the Ecological Cities Project (EC). Among many related activities, we organized a series of regional EC conferences with local collaborators in Boston, Columbia (South Carolina), Milwaukee, and with our visiting Turkish colleague Azime Tezer in Istanbul.

In June 2002, our road show came to New York City. With support from the Lincoln Institute of Land Policy, the first Humane Metropolis Symposium was held at the New York University Law School Auditorium on Washington Square, a few blocks from where I grew up. The program honored the work of the late Holly Whyte through personal testimonials and reports on work in progress reflecting his interest in urban people, places, and greenspaces. The term "humane metropolis" was my update on *The Exploding Metropolis,* the heretical 1958 book by Whyte, Jane Jacobs, and others that skewered postwar urban policies (see chapters 3 and 4). The point is: now that the metropolis has "exploded," how can we make it more liveable and humane?

The Humane Metropolis book, as I mentioned earlier, served as a template for three more Humane Metropolis conferences in Pittsburgh (2007), Riverside, California (2008), and Baltimore (2009), all organized in collaboration with the Lincoln Institute of Land Policy. In a similar vein, as a senior fellow with the CUNY Institute of Sustainable Cities, I conducted a study of the New York City waterfront culminating in a series of public panels in 2010 titled *Turning the Tide: New York's Waterfront in Transition.*

The several dozen invited speakers at all these events were primarily "movers and shakers" (not technocrats) striving to upgrade the condition of people, places, and nature in their respective localities. These and many other folks I have encountered at conferences, speaking appearances, and personal travel are the source of much of what is related in this book, especially Part III. They and their predecessors dating back to "the Janes" (Addams and Jacobs) and other past activists have challenged the smug certainties of the old order, releasing the energies, knowledge, talents, and resources of grassroots leaders to gain traction in pursuing their respective urban visions.

Among the countless individuals whose collective efforts comprise the "compost" of humane urbanism, I am particularly indebted to: Mike Houck and Steve Johnson in Portland, Oregon; Jane and Richard Block in Riverside, California; Andy Lipkis, Lewis MacAdams, and Jennifer Wolch in Los Angeles; Will Allen in Milwaukee; Chris DeSousa in Milwaukee and Toronto; David Crossley, Terry Hershey, and Anne Olson in Houston; Gail Thomas in Dallas; Jerry Adelmann,

Patsy Benveniste, Joel Bookman, Mark Bouman, Ron Engel, Suzanne Malec-McKenna, Alex Polikoff, Melinda Pruett-Jones, Laurel Ross, Ed Uhler, and Susan Yanun in Chicago; Joy Abbott in Pittsburgh; Larry Aurbach, Steve Coleman, Peter Harnik, and Betsy Otto in Washington, DC; Drew Becher, Eugenie Birch, Charles Carmalt, Howard Neukrug, and Maitrey Roy in Philadelphia; Bob Zimmerman, Newton, Massachusetts; Anne Whiston Spirn in Cambridge and Philadelphia; Jackie Carrera, Guy Hager, and Ed Orser in Baltimore; Margaret Carmalt and Colleen Murphy-Dunning in New Haven; Mary Rickel Pelletier, Hartford; Harvey Flad in Poughkeepsie; and Ann Buttenwieser, Roland Lewis, Philip Lopate, Peter Mullin, Regina Myer, Rob Pirani, Robin Simmen, Cortney Worrall, and Bob Yaro in New York City. I especially thank Armando Carbonell at the Lincoln Institute for his encouragement and support for the *Humane Metropolis* book and related conferences. I appreciate thoughtful reviews of earlier drafts of this manuscript by Catherine Tumber and Alex Marshall. Thanks to Nancy Denig for her poem in the introduction. And I owe special thanks to Laurin Sievert for all her help in developing the Ecological Cities Project and to Kathleen Lafferty, faithful editorial gadfly and dear friend. And once again, I salute the helpful and efficient staff of the University of Massachusetts Press, especially Bruce Wilcox, Clark Dougan, Carol Betsch, Jack Harrison, and Karen Fisk. Mary Bagg did an excellent job copyediting the manuscript on a tight schedule.

RECLAIMING AMERICAN CITIES

Introduction:

A Train Journey into the Past and Future

Life must be lived amidst that which was made before. Every landscape is an accumulation. The past endures.
—DONALD W. MEINIG, *The Interpretation of Ordinary Landscapes,* 1979

Six a.m. is an ungodly hour to begin a train journey (or a book). My ride to New York City one fine summer morning begins at Springfield, Massachusetts, the northern terminus of Amtrak's Connecticut River valley line. Springfield in its heyday was a major rail hub with some two hundred trains daily and hundreds of passengers thronging the waiting hall and platforms.[1] Today, Springfield's station is a barebones trackside ticket counter and seating area, as welcoming as a jury waiting room. We file out to the two-car train whose diesel locomotive (the line is not yet electrified) belches exhaust into the morning mist. Untouched by the computer revolution, conductors punch paper tickets and off we go at a blistering crawl.

A humdrum morning train ride might normally not be regarded as an opportunity to reflect on the imprint of humanity upon the world around us, and indeed most of my fellow passengers are soon sensibly asleep. But given a window seat and a sprinkling of imagination, the view from the train offers a panorama of the Connecticut Valley's settlement history and a glimpse of its future. As the environmentalist Bill McKibben pungently described this very route: "Along the way, the river cuts through many strata of biology and history and economy—small farms, old mill towns dating back to the zenith of New England's industrial revolution, small cities slowly rotting."[2] The trip just now beginning will offer glimpses of such earlier stages of the "enduring past" (in Meinig's phrase) as well as hints of a new era for those post-rotting small cities, a further layer of accumulating landscapes which suggests that they may fare better in the century now in progress than during the final decades of its predecessor.

After creeping past derelict freight facilities and beneath an interstate highway, the train, like the taxi in Woody Allen's *Midnight in Paris,* warps back in time. For ten minutes, we follow the eastern shore of New England's master stream running free and frothy over rocks and around islets in sparkling sunshine—the combined flows of tributaries from Canada to the north, the White Mountains to the east, and the Green Mountains and Berkshires to the west. This river and its glacier-scoured valley have sustained humans since time immemorial. Native peoples left their imprint in the many settlement sites, trail routes, and patches of forest clearance that greeted the first European settlers.[3] Lured by the fertile lowlands, upland forests, furry and edible wildlife, and teeming fisheries, a chain of seventeenth-century colonial outposts rapidly ascended the valley, reaching the site of Springfield in 1636.[4] The transition from Native to European occupance of the valley constituted an epic demonstration of Meinig's "landscape accumulation." The historian William Cronon has described the clash of cultures: "When the Europeans first came to New England, they found a world which had been home to Indian peoples for over 10,000 years. But the way Indians had chosen to inhabit that world posed a paradox almost from the start for Europeans accustomed to other ways of interacting with the environment. Many were struck by what seemed to them the poverty of Indians who lived in the midst of a landscape endowed so astonishingly with abundance."[5]

Whereas Natives treated the land as a "commons" whose bounty (crops, wildlife, fish, firewood, berries, etc.) was shared as needed for survival, the Europeans imported their notions of "ownership" with the concomitants of legal title, marketability, fencing, and pursuit of private economic profit. The English also imported their political geography of counties and towns, thereby dividing the "wilderness" into discrete administrative and economic units that feistily endured to the present day. Vestiges of the valley's agrarian past are glimpsed from the train: stone walls, weathered barns and silos, farmhouses of wood or stone, gleaming Congregational churches—the "enduring past" amid the cluttered present of fast food outlets, billboards, municipal buildings, and parking lots.

The train crosses the river in view of the Enfield Dam, a crumbling relic of industrial New England, built in 1827 at the river's first rapids above tidewater. Ingeniously, the dam was equipped with a dual canal system to power mills and pass river traffic beyond the rapids. Its current decrepitude benefits migratory Atlantic salmon and shad long blocked from the spawning grounds upstream. A rival overland canal that once connected New Haven and Northampton, Massachusetts, was converted to a rail corridor in the 1840s, which is now again being transformed into the eighty-four-mile Farmington Canal Heritage Bikeway: more landscape accumulation.

We pull into Hartford—the wealthiest city in the United States after the Civil War, a former industrial powerhouse, and home of Mark Twain, Louisa May Alcott, and Harriet Beecher Stowe. By 2002, it was "the most destitute 17 square miles in the nation's wealthiest state, and a city where 30 percent of its residents live in poverty."[6] Hartford from the train window is a tangle of overhead expressways described by Catherine Tumber in *Small, Gritty, and Green* as a "tectonic mash-up . . . right in the middle of downtown. The multilevel six-to-eight-lane tangle of concrete, asphalt, and steel stretches more than three miles."[7]

But Hartford and its regional peers in Massachusetts, such as Springfield, Worcester, and Holyoke, are like aging dowagers endowed with treasures from earlier times (parks, colleges, libraries, museums) amid the wreckage of decline. Today these "legacy cities" are inspiring homegrown efforts to make them greener and more habitable,[8] such as Mary Rickel Pelletier's long quest as director of a watershed initiative to revive the north branch of Hartford's Park River, ecologically and socially. Upstream, the old "Paper City" of Holyoke is the scene of a Hispanic-led urban farming and economic development program: Nuestras Raices.[9] Our train stop in New Haven amid the detritus of failed urban despotism of the 1960s is the venue of the Urban Resources Initiative, a remarkable collaboration between Yale University, the city, and its neighborhoods. (See chapter 10 for case studies of the Holyoke and New Haven initiatives.)

Now zipping along the Connecticut shoreline on the electrified mainline toward New York, the train view alternates between scenes of incredible squalor and of incredible wealth. The blackened, graffiti-strewn, windowless hulks of former industrial plants in Bridgeport and Norwalk are succeeded by the Lexus dealers, upscale shopping centers, and weirdly pretentious McMansions of Greenwich and Darien. The train stops at Stamford with its glitzy corporate headquarters and the usual Marriott hotel, convention center, and other trappings of a classic "edge city."[10]

Entering the Bronx, New York City's northernmost borough, we sweep past Co-op City, a state-subsidized complex of fifteen thousand affordable apartments, offices, and retail shopping—a legacy of the 1960s quest to provide an in-city halfway

---

Battleground

Up and down
the railway corridor,
I grieve
the blinded
factory windows,
rusted metal trusses
and brave green tufts
in vacant parking lots
scrolling by
my window.
Romanced
as I am by trains,
I see these scenes
as something
heroic,
ground down
in defeat, yet
not without a battle—
nor hope.

Nancy Denig, December 18, 2010

station for "white flight" between the older urban neighborhoods and the more expensive suburbs beyond.[11] Barely glimpsed from the speeding train are blocks of row houses and low-rise apartments created by a massive public-private effort to revive the impoverished and crime-ridden South Bronx during the 1990s.[12]

The train swoops past whatever the rest of the city has banished to the South Bronx—an immense food distribution center, a sewage treatment plant, a power generating station, Rikers Island prison, another prison floating on barges. Triborough Bridge, the hub and cash cow of the Robert Moses era, lies to the right, and his Randall Island park complex to the left of our elevated track. But even this poor and long-oppressed sinkhole of the metropolis is undergoing a gradual revitalization through the joint efforts of a network of nonprofit organizations including Sustainable South Bronx, the Bronx River Alliance, and Bette Midler's New York Restoration Project, with strong support from city agencies under Mayor Michael Bloomberg's *PlaNYC*.[13] This collaboration is gradually achieving "an unexpected renaissance . . . along the south end of the Bronx River," the *New York Times* reported. "Park by park a patchwork of green spaces has been taking shape, the consequence of decades of grinding, grass-roots, community-driven efforts."[14]

The train on its elevated right-of-way soars across a sea of postwar bungalows in Queens, now home to thousands of Asian and Eastern European households. With a glimpse of the last best hope of internationalism, the United Nations headquarters, we plunge into the century-old East River Tunnel and arrive at New York's despised Penn Station, the squalid successor to the Beaux-Arts masterpiece whose demolition in 1963 gave rise to the historic architecture protection movement.[15]

Having traveled from the empty streets of early morning Springfield to the madhouse of mid-day Manhattan—and through time from the seventeenth into the twenty-first century—the obvious question is What now? As Scrooge pleaded for a chance at redemption after being shown the Ghosts of Christmas Past, Present, and Future, we may rightly be appalled at the specter of what went wrong with city building and rebuilding during the last century and fear what lies ahead. But the accumulating landscapes glimpsed or visualized on our train journey suggest that our collective urban future may in certain respects be an improvement on the urban past. *Or at least it will be shaped in important ways by the people who inhabit ordinary urban and suburban communities, rather than by the cultural, economic, and technocratic elite with their "one-size-fits-all" solutions that often caused more problems than they solved.*

The Plan of the Book

This book joins a crowded field. Among recent urban titles to be found in the nation's dwindling full-service bookstores, online, and on the cluttered floor of my office are Anthony Flint's *This Land* (2006); Harry Wiland and Dale Bell's,

Edens Lost and Found (2006); Neal Peirce and Curtis W. Johnson's *Century of the City* (2008); Jeb Brugmann's *Welcome to the Urban Revolution* (2009); Joel Kotkin's *The Next Hundred Million* (2010); *Wild in the City,* edited by Michael C. Houck and M. J. Cody (2011); Edward Glaeser's *The Triumph of the City* (2011); John D. Kasarda and Greg Lindsay's *Aerotropolis* (2011); Timothy Beatley's, *Biophilic Cities* (2011); Catherine Tumber's *Small, Gritty, and Green* (2012); and *Rebuilding America's Legacy Cities,* edited by Alan Mallach (2012), among the many others I cite throughout.

This book departs from the urban sustainability pack in three respects. First, it is not another "how to" guidebook to urban sustainability. It is less concerned with the details of specific strategies and techniques (e.g., green buildings, low-impact development (LID), transit-oriented development (TOD), urban farming and forestry, ecological restoration) than with the balance of influence between top-down and bottom-up sources of urban priorities and initiatives; namely, *who decides what is important, for whose benefit, how to achieve it, and at whose expense?* Second, it is less concerned with "downtown" and its "edge city" clones than with the habitability of ordinary urban and suburban communities. Third, this book takes a historical approach, tracing the evolution of urban policies and perspectives over the past century rather than treating urban sustainability as appearing full-blown "from the head of Zeus" with the advent of Smart Growth and New Urbanism in the 1990s.

Urban policies of the past century have been dominated by confidence in the infallibility of expert-driven solutions—architects in the earlier decades, developers and public officials in the postwar decades, and assorted policy wonks to the present time. Though challenged by the likes of Jane Addams in the early 1900s, Jane Jacobs, William H. Whyte, and Ian McHarg in the 1950s and 1960s, and Anne Whiston Spirn, James Howard Kunstler, and Tony Hiss in the 1980s and 1990s, urban policies have been essentially top-down, one-size-fits-all, and expert-driven. Cities and suburbs have been the supposed beneficiaries (often as willing collaborators) of policies, programs, and priorities developed on their behalf by higher levels of government, corporations, consultants, and the nongovernmental sector. For decades after World War II, the conventional wisdom assumed that *the future belongs to downtown, the suburbs, the white establishment, and the automobile.*

The book's time span is divided into three chronological periods. Part I, "The Patrician Decades, 1900–1940," revisits the Progressive Era of "good works," both architectural and social, as reflected in the competing agendas of the City Beautiful and settlement house movements, and their uneasy synthesis in the city and regional planning movement. These were the golden years of exuberant city center growth (especially upward) and well-meaning philanthropy, thanks to the dominance of the corporate, financial, and cultural elite. This power structure was challenged primarily by dedicated young women inspired by Jane Addams, founder of Hull House and its diaspora across the nation. Their populist struggle against the downtown elite on behalf of housing, working conditions, and health of the poor gained influence within the New Deal, as reflected in the appointment by President Franklin D. Roosevelt of Frances Perkins—a former Hull House volunteer—as secretary of labor and first female cabinet member. The Housing Act of 1937 resulted from the advocacy of Catherine Bauer and other settlement house partisans. Postwar America, however, would be more strongly shaped by the worldviews of two uber-architects, Frank Lloyd Wright and Le Corbusier—as popularized in widely acclaimed exhibits at the 1939 New York World's Fair.

"The Technocrat Decades, 1945–1990" (Part II) explores the fateful shift in influence over the future of cities from the civic and urban design elite to federal, state, and local bureaucrats, influenced by a new generation of highway engineers, modernist architects, planners, and public administrators personified by New York's Robert Moses. Chapters 3 and 4 outline the apartheid postwar national policy "engines" that respectively powered central city "renewal" and suburban sprawl under the direction of the technocracy. In reaction to the social and environmental impacts of these policies, the "struggle for people, place, and nature" evolved in new directions between the 1950s and the 1980s. These included demands for civil rights and housing reforms, including the epic *Gautreaux* public housing litigation, Great Society reforms, and legal challenges to exclusionary zoning; local battles to save neighborhoods and parks from urban highway construction, as catalyzed by Jane Jacobs and others; and efforts to save threatened "Last Landscapes" from industrial and urban development.

"The (More) Humane Decades, 1990–Present" (Part III) traces the advent of urban sustainability in its many flavors, the spread of the Smart Growth/New Urbanist movement, and the maturing of community advocacy and grassroots empowerment. After a century of top-down, expert-driven, "take your medicine" nostrums decreed by patricians and technocrats, local community organizations, special interest groups, and private citizens began to make progress in defending themselves, their neighborhoods, and their local environments. The selection of 1990 as a divide between Parts II and III is subjective—certainly the preceding three decades witnessed many portents of more humane urban policies. But as I argue in chapter 7, the 1990s—spanning the last two years of the George H. W. Bush administration and both Clinton terms—yielded a remarkable harvest of new tools such as the Americans with Disabilities Act of 1990, new urban strategies under the banner of Smart Growth, and revitalization of earlier laws such as the Community Reinvestment Act of 1977. Collectively the "garden of acronyms" that accumulated (like the landscapes they influenced) helped to empower community-driven humane urbanism with new legal authorities and sources of funds.

The Meaning of Humane Urbanism

"Humane urbanism" refers to efforts to make cities and suburbs:

- Greener (more sustainable, reduced ecological footprint, restoration of habitats, etc.).
- Healthier and safer (promotion of healthy lifestyles, reduction of natural hazard risks).
- More equitable and multicultural.
- More efficient (use of energy, water, materials, time, etc.).
- More people-friendly and fun.[16]

As compared with the macro-urbanism of the top-down decades, humane urbanism is more concerned with the revival of micro-places like vacant lots, empty storefronts, and dead shopping centers; litter-strewn parks and stream corridors; abandoned rail rights of way; the incidental patch of native forest, prairie, bog, or a single landmark tree; drab row housing, closed schools, decrepit historic structures; and so on. Under the growing influence of humane urbanism, cities have become regarded not only as economic engines dominated by downtown and the corporate/technocrat/cultural elite, but also as places to care about, live in, and enjoy—whose inhabitants play key roles in determining the futures of their urban homes. Corresponding to the micro-scale of opportunities for humane urbanism is a new generation of "micro-actors," Some of these may be experienced local or regional nongovernmental organizations (NGOs) concerned with particular issues such as affordable housing, schools, health, disabilities, public transit, or pedestrian and cycling convenience and safety. Others are more informal, multi-issue neighborhood and block coalitions, "Friends of" groups, ad hoc alliances and partnerships, and garden-variety volunteer networks. Humane urbanism is not confined to large cities with vast resources to support community initiatives. It also may thrive in smaller cities like the former industrial and mill towns of New England, New York State, and the upper Midwest.[17]

The Smart Growth / New Urbanist movement, as I discuss in chapter 7, has helped to foster humane urbanism (whether its polemicists are aware of it or not) by encouraging new ways of thinking about existing urban places at various scales. Project for Public Spaces, for instance, has applied the people-friendly design ideas of William H. Whyte (as refined by New Urbanists) to the replanning of hundreds of small urban public spaces over the past three decades. But for the vast expanses of metropolitan America that are not targets of investment opportunity any time soon (dubbed "midopolis" by Joel Kotkin[18]), the "humane decades" have spawned a "garden of acronyms"—new or recycled legal devices and funding strategies to make metropolitan America more humane in diverse ways. In the final three chapters I sample selected applications of humane urbanism as reflected in "New Age" downtown parks (chapter 8), urban watershed restoration (chapter 9), and "down to earth" programs to grow food, trees, and community in diverse cities and neighborhoods around the country (chapter 10).

Of course, not everyone is on the same train (if there is still a train to be on). In a throwback to the heyday of elitist planning, two books I already mentioned above, both published in 2011—*Aerotropolis: The Way We'll Live Next* by John D. Kasarda and Greg Lindsay, and Edward Glaeser's *Triumph of the City: How Our Greatest Invention Makes Us Richer, Smarter, Greener, Healthier, and Happier*—view cities first and foremost as the domain of the plutocracy, namely, those who are or hope to be among the top few percent of Americans (and their global counterparts) who earn and own as much as everyone else combined.

In contrast to those at the apex of the social pyramid, most people living in metropolitan America (some 240 million) are not rich, powerful, or patrons of rooftop restaurants and health clubs. They scrape along somewhere on the spectrum between abject poverty and reasonable financial security. As the economist Don Peck observed the *Atlantic* (also in 2011): "Arguably the most important economic trend in the United States over the past couple of generations has been the ever more distinct sorting of Americans into winners and losers, and the slow hollowing out of the middle class."[19] That growing wealth divide is real and abhorrent, and those who pander to the plutocracy make no apologies. But it is rather stark to write off everyone outside the magic circle of the wealthy and privileged as "losers." Life offers many sources of satisfaction other than wealth alone (if that is, in fact, a source of satisfaction). "Winning" also may be defined in terms of family, friendships, spirituality, public service, artistic pursuits, sports and outdoor recreation, and other life interests.

And contrary to the "urban triumphalists," cities cannot be declared "winners" solely in terms of how high their skyscrapers reach and how many international flights operate from their airports. Nicholas Lemann observed such an inclination in the *New Yorker*: "Urban strategists have a penchant for magically simple explanations, but Kasarda takes it to an extreme: for him, air freight is the core element of civilization." More broadly, Lemann rejects "one-size-fits-all" solutions as irrelevant to the needs and aspirations of most urban and suburban inhabitants in the twenty-first century: "Many people are unwilling to participate in a social compact based on perpetual motion. Masters of the new economy, social visionaries, and tongue-studded app developers figure large in the imagination of urban theorists these days, but *most people are looking for something pretty mundane*: a neighborhood, a patch of ground, a measure of peace and security, a family, status, dignity."[20]

In the same vein, Paul Grogan and Tony Proscio laud the grassroots results of community development corporations (as of 2000): "How implausible it was that a ragtag bunch of neighbors, in communities deserted by most mainstream institutions, could make such a difference! It was *the modesty of their ambition*—not to save the nation or the world, but a house or a block or a neighborhood—the very concreteness, the localness of what they set out to do—that now highlights the stunning scope of their achievement."[21]

Macro-planning, for better or worse, still thrives in certain settings, such as Atlantic Yards in Brooklyn, the Anacostia Waterfront Initiative in Washington, DC, the proposed BeltLine in Atlanta, and the Gateway Arch riverfront enhancements in St. Louis. But countless micro-scale improvement projects erupt in less auspicious locations, for example, the Marvin Gaye Greenway in the nation's capital (which I discuss in chapter 10), a world apart from the new baseball park and other mega-projects just downstream on the Anacostia. Like wildflowers sprouting from the cracks of abandoned parking lots, humane urbanist initiatives are largely spontaneous and self-sustaining. They make their surroundings more bearable and local inhabitants more connected to one another and to natural and cultural phenomena in their midst. As the son of a natural history writer, I suggest an earthy metaphor for humane urbanism: that it's akin to the soil mantle as a living habitat where all manner of plants, insects, worms, bacteria, and fungi thrive through mutual symbiosis.

Grogan and Proscio's "ragtag bunches of neighbors" are now more frequently writing the script, and the well-paid consultants and bureaucrats, if involved at all, must get over being the stars of the show and settle for being the supporting cast. Some of the nation's elected municipal leaders like Chicago's Richard M. Daley, New York's Michael R. Bloomberg, Washington's Anthony Williams, and Los Angeles's Antonio Ramón Villaraigosa have embraced and facilitated the emergence of community and stakeholder empowerment, while helping to re-align federal and state priorities to facilitate rather than dictate local priorities and outcomes.

Admittedly, this book is written in the context of "A Growing Gloom for States and Cities," to borrow the phrase from the title of a *New York Times* editorial on August 13, 2011. While the worsening fiscal crisis facing government at all levels certainly limits funding from those sources, the resilience of investment markets during the recession has apparently allowed the foundation sector to maintain and sometimes enlarge its contribution to locally initiated projects. Key regional nongovernmental organizations such as Bette Midler's New York Restoration Project, Openlands in Chicago, the Trinity Trust in Dallas, and the Local Initiative Support Corporation (LISC) in various cities continue to thrive. And volunteerism is free!

In any event, there is no turning back: the glorious illusions of the past century—escapism (to suburbs, Sun Belt, or "nerdistans"),[22] the primacy of rich white males, unlimited water and energy, and the dominance of the automobile—all have proven fallacious. We have to work with what we have, and shelve one-size-fits-all dogmas.

The following list excerpted from a 2007 article by Jennifer Wolch, a geographer then at the University of Southern California, now at Berkeley, enumerates the preeminent challenges for twenty-first-century urban practice and research:

1. To reweave the urban fabric as a vital green matrix, to conserve and restore habitat and watersheds.
2. To transition toward more sustainable patterns of urban production and consumption.
3. To recast the rights and obligations of citizenship . . . as a means to challenging hegemonic structures and institutions, promoting social and ecological justice, and moving toward greener urban worlds.[23]

In summary, this book examines both sides of this historic sea change: the well-meaning but often oppressive footprint of the "hegemonic" past and the promising but still tentative first steps toward achieving more livable and enduring communities today. For better or worse, our metropolitan complex is largely built; now we have to make it as bearable, sustainable, and humane as humanly possible.

PART I

The Patrician Decades, 1900–1940

The first third of the twentieth century—bracketed by the inaugurations of
President Theodore Roosevelt in 1901 and his distant cousin Franklin D. Roosevelt
in 1933—was the golden age of the American city. It was the heyday of tall slen-
der "skyscrapers," Model A Fords and Stutz-Bearcats, high-speed and luxurious
inter-city railroads, convenient and affordable commuter rail service, the spread
of national radio networks, the rise of big city professional sports, and the convenience of buying and selling stocks via "wire" or telephone. America before it
entered World War I in 1917 was a country of smug confidence bred of growing
corporate wealth and overseas imperialism. It was also a time of unfettered laissez-
faire, widening disparities in wealth, pervasive political corruption, and abject living
conditions for the poor. As a result reformist efforts sprang from various strands of
progressivism, including urban beautification, clean government, anti-trust, natural
resource conservation, and, most enduringly, the settlement house movement and
its offshoots concerning sanitation and public health, housing, playgrounds, child
labor reform, minimum wage, unionization, workplace safety, and public education.

After the war and a raging influenza pandemic, the nation's cities rebounded as
the stage sets for the Roaring 20s and the Jazz Age. Prohibition, lasting from 1919
until 1933, fostered the titillating culture of the speakeasy, along with organized
crime and mob violence epitomized in Al Capone's 1929 St. Valentine's Day Mas-
sacre. (The legendary violence of the era would be a staple of late twentieth-century
period films like *Bonnie and Clyde, The Sting, The Untouchables,* and *Road to Perdi-
tion*). This was the era memorialized in F. Scott Fitzgerald's *The Great Gatsby* and
The Beautiful and Damned, Sinclair Lewis's *Main Street* and *Babbitt,* Hemingway's *A
Farewell to Arms,* Gershwin's *Rhapsody in Blue.*

The twenties was a time of instant heroes like Babe Ruth ("the King of Swat")
and Charles A. Lindbergh ("Lucky Lindy"). It was an era of religious and ethnic
witch hunts most famously reflected in the Scopes Monkey Trial in 1925 and the
chilling executions of Sacco and Vanzetti in 1927. Corporate profits and influence
flourished under the complacent Republican administrations of Harding, Coolidge,
and Hoover, which collectively spanned 1921–1933. Big city machines and bosses
such as Tammany Hall in New York, Mayor James Michael Curley in Boston, and
"Big Bill" Thompson in Chicago ran their cities like medieval fiefdoms, with anyone
seeking political favors paying the necessary kickback. Stock prices soared like the
102-floor Empire State Building, driven by an unshakable faith in a permanent state
of peace and prosperity. In Scott Fitzgerald's words, it was an "era of wonderful
nonsense." It also proved, of course, to be a house of cards.

Chapter 1

American Cities in 1900: A Patchwork of Silk and Rags

We must face the inevitable. The new civilization is certain to be urban, and the problem of the twentieth century will be the city.
—JOSIAH STRONG, *The Twentieth Century City*, 1898

The dawn of the twentieth century was the sunset for America as a nation of farmers, villages, and mill towns. As reflected in the populist revolt of the mid-1890s, the people of the nation's heartland felt threatened by forces beyond their control: Big Capital, Big Industry, and Big Cities. (Big Government would be the bête noire of the latter-day populism known as the Tea Party.) Although the nation's population in 1900 was still more rural than urban,[1] the balance was changing rapidly with rising birth rates and improved sanitation in cities, and with immigrants from the countryside and from abroad in search of new opportunities, freedom from rural toil, and "bright lights" of city life.

Suddenly, a Nation of Cities

The 1920 Census disclosed that the nation's urban-rural balance had reversed: for the first time, people living in cities now outnumbered those still in small towns or the countryside. The Age of Jefferson and Jackson was yielding to the Age of the Roosevelts. That demographic milestone could be viewed as the logical sequel to Frederick Jackson Turner's thesis that the "closing of the frontier," as reported in the 1890 Census, spelled the end of pioneer settlement and its role as a liberating and democratizing influence in American civilization. Now, it could be argued, cities were replacing the frontier as the "promised land" to the restless, the lonely, and the dispossessed who swarmed to them by coach, rail, ship, steamboat, or foot. Like the western frontier, cities were hazardous territory for newcomers—rough and tumble, rife with street crime, inebriation, prostitution, disease, and even murder, as chronicled by Erik Larson in *The Devil in the White City*, set in Chicago.[2] As the urban historian Jon C. Teaford has written: "[In 1900] the farm no longer provided sufficient employment or excitement for the world's masses, so they turned to the urban hubs of manufacturing, commerce, culture, and amusement. Migrants from the peasant villages of Europe and Asia, and from the declining farms of New England and the Midwest, converged on the large metropolises, a motley collection of humanity who together committed their fate to the American city."[3]

Parallels between Turner's "closing of the frontier" thesis and the migration to cities, however, are subject to a major qualification. Reflecting the populist movement of his own time—and presciently anticipating today's Tea Party libertarianism—Turner warned that freedom and individuality nurtured by pioneer settlement could get out of hand, leading to "antipathy to control, and particularly to any direct control. The tax-gatherer is viewed as a representative of oppression. . . . Democracy born of free land, . . . intolerant of administrative experience and education, and pressing individual liberty beyond its proper bounds, has its dangers as well as its benefits."[4] Resistance to limits of personal freedom in the frontier context was part of the mythic narrative of the "Wild West," but such resistance in a predominately urban and metropolitan context is potentially catastrophic.

The View from Downtown: Bigger Is Better

The twentieth century opened with many portents, both promising and foreboding, for the future of America and its cities. The nation was rebounding from the depression of the mid-1890s and its wave of labor and political unrest. With the brief but glorious (according to the Hearst newspapers at least) Spanish-American War just won, the United States was basking in a newfound imperial self-confidence with the acquisition of Cuba and the Philippine Islands. The nation's white, conservative, male elite were amassing wealth, immense for a few and considerable for many others. According to the historian Edmund Morris: "The United States was the most energetic of nations. She had long been the most richly endowed. This first year of the new century found her worth twenty-five billion dollars more than her nearest rival, Great Britain, with a gross national product more than twice that of Germany and Russia. The United States was already so rich in goods and services that she was more self-sustaining than any industrial power in history."[5]

In a sense, the urban history of twentieth-century America began on the windy shore of Lake Michigan in Chicago with the World's Columbian Exposition of 1893, a landmark among world's fairs. The fair reflected the brash energy and confidence of a city swollen from 30,000 to nearly 1.7 million over the second half of the century, even despite its Great Fire in 1871. The exposition established that precocious city (in its own opinion at least) as a "world class" rival to New York, Paris, and London as a center of business, culture, and civic pride. The fair looked to the future as an awesome demonstration of electricity's capacity to power and illuminate its exhibits and grounds ("The White City;" see fig. 2.1). It also looked back to antiquity in its Beaux-Arts neoclassicism then in vogue—imitating Paris's Place de la Concorde and the palace of Versailles on the shore of Lake Michigan.

Fig. 1.1. Traffic congestion in downtown Chicago Loop, 1909. Courtesy of the Chicago History Museum. Neg. ICH1-04191; photographer unknown.

The fair would influence civic architecture and planning through the City Beautiful movement well into the new century.

While the fair attracted millions to Chicago's South Side lakefront, an architectural revolution having nothing to do with nymphs, fountains, and Greek porticoes was unfolding in the city's "Loop"—the central business district enclosed by elevated tracks and choked with traffic and pedestrians (fig. 1.1). Until the 1890s, multistory buildings were supported by their exterior masonry walls; the higher the building, the thicker the weight-bearing walls had to be, which consumed much valuable, interior floor area at street level. Late in the century, a new generation of gifted Chicago architects, including Louis Sullivan, Dankmar Adler, William Le Baron Jenney, Daniel H. Burnham, and the legendary builder George A. Fuller, began to experiment with the use of steel girders to support tall buildings, with exterior walls serving only to enclose the interior from weather. The Monadnock Building of 1893, one of the city's most treasured landmarks (and where I once worked) was transitional: the earlier sixteen-story section was the highest weight-bearing wall structure ever built; steel framework would support a later expansion. Like railroads eclipsing mule-drawn canal boats, the steel frame replaced the weight-bearing wall once and for all. Steel frames, and the invention of faster elevators, fostered a skyscraper race between Chicago and New York in the early 1900s that resumed in the late twentieth century.[6]

A less visible but city-saving Chicago achievement at the turn of the century was the completion in 1901 of the Chicago Ship and Sanitary Canal. This huge engineering milestone reversed the flow of the sluggish and foul Chicago River to convey the city's raw sewage away from, rather than toward Lake Michigan, the city's primary drinking water source.[7] While protecting Chicagoans from drinking their own wastes, the canal transported the city's raw effluent inland to the westward-flowing Des Plaines and Illinois rivers, until sewage treatment was finally provided after World War II. It also replaced the old Illinois & Michigan Canal as a navigation link between Lake Michigan and the inland waterways of the Midwest.

New York City, meanwhile, could shrug off Chicago's pretensions with its own talent, wealth, and capacity for innovation. As the new century arrived, New York could boast of a long series of milestones in its evolution from colonial port to metropolis. By 1900, the city had constructed and enlarged its Croton River water supply project that impounded, conveyed, and distributed potable water throughout the city from a Hudson River tributary forty miles north of the city. Boston and many other cities applied the Croton model in developing their own upland sources of freshwater that flowed by gravity to the user region. Later expanded tenfold with new reservoirs west of the Hudson River, New York's water supply system remains a world gold standard.[8]

New York's Central Park was another product of the city's foresighted leadership in the mid-nineteenth century. Designed and constructed by the soon-to-be legendary landscape architect Frederick Law Olmsted and his partner Calvert Vaux, the park would be the exemplar for a generation of urban parks across the country from Boston to San Francisco. Among its signature design elements were contrasting spaces devoted to open meadows and woodlands, gardens and picturesque patches of "wildness," several artificial water bodies, and the separation of internal circulation routes for carriages, horseback riding, and strolling (and much later, cycling, running, and skateboarding). The park was designed in coordination with the Croton water project to incorporate a receiving reservoir, which was later converted to open play fields.

The stunning Brooklyn Bridge, which opened in 1888, was designed and built by the German-born father and son John and Washington Roebling. The bridge was suspended from fifteen-inch diameter cables containing hundreds of wire strands spun in place using new technology invented by the elder Roebling. The bridge's great masonry towers, the two highest structures in the city at the time, rested on bedrock beneath the East River using caissons dewatered with compressed air to give the laborers access to the work sites.[9]

The bridge in turn fostered a political triumph: the creation of the five-borough New York City. As a physical link between the still-independent cities of New York (Manhattan) and Brooklyn, the bridge helped to strengthen proposals to unite those cities.[10] In 1898, they, along with Queens, the Bronx, and Staten Island, were politically consolidated by popular referendum into Greater New York City, creating a metropolis that "over three million strong, over three hundred square miles huge, larger than Paris, gaining on London, . . . was ready to face the twentieth century."[11] In 1910, New York was the second-most populous city in the world behind London. (Today, metropolitan New York is the last city of the Industrial Era remaining among the world's ten largest urban regions, which otherwise include megacities of East and South Asia and Latin America.)[12]

Even as the city expanded geographically and demographically, its downtown Manhattan business district steadily became more intensely developed and congested. Daniel H. Burnham's famous twenty-story Flatiron Building opened in 1902 at the wedge-shaped intersection of Broadway and Fifth Avenue as "a bridge between the picturesque world of gaslight New York and the skyscrapers of the aggressive 20th century commercial market"[13] (fig. 1.2). In the same year and ten blocks to the north, Macy's opened what was claimed to be "the World's Largest Store" at Herald Square where Broadway intersects Sixth Avenue. Also in 1902, the New York Central Railroad inaugurated its luxurious Twentieth Century Limited express trains between New York and Chicago, with connecting service to and from Boston.[14] The rival Pennsylvania Railroad launched its competing Broadway Limited Chicago service a few years later. Both railroads constructed mammoth gateways to the metropolis: Pennsylvania Station (1910; demolished 1964) and Grand Central Terminal (1913). The advent of electrified trains allowed tracks serving these stations to be buried in tunnels, leaving the overhead "air space" available for real estate development: Park Avenue's swank apartment buildings arose above the New York Central tracks north of Grand Central.

The nation's capital in 1900 had reached a population of 278,000, a nearly fivefold increase since 1860. The city had long outgrown its original 1791 plan by the French engineer Pierre L'Enfant, which sketched the basic outline of streets, parks, and seats of government (White House and Capitol). In the late 1800s, the filling of tidal flats bordering the Potomac River presented the city (and nation) with the opportunity to redesign its central area. Congress as the local government for the District of Columbia established the Senate Parks Commission (McMillan Commission) to prepare a new plan for the city's core area and outlying

Fig. 1.2. Edward Steichen, "The Flatiron," 1909. Permissions © The Estate of Edward Steichen. Image copyright © The Metropolitan Museum of Art. Image source: Art Resource, NY.

park systems.[15] The resulting public exhibition and plan would guide the future growth of the city and contribute to the expanding influence of the City Beautiful movement.

Meanwhile, turn-of-the-century San Francisco, before the catastrophic earthquake and fire that engulfed it in 1906, was becoming a powerful urban center, according to an ecstatic booster: "The great triangle of the Pacific is destined to have its lines drawn between Hong Kong, Sydney, and San Francisco. Of these three ports, Hong Kong will have China behind it, Sydney, Europe, and San Francisco, America; and with America for a backing, San Francisco can challenge the world in the strife for commercial supremacy."[16]

The 1906 San Francisco earthquake caused one of the world's worst non-war conflagrations due to ruptured gas lines, prevalent wood construction, and inadequate water supply. The area burned was six times larger than the extent of the London fire in 1666 and half again as much territory as the 1871 Chicago fire. Approximately five hundred people died in the catastrophe, three-fifths of the city's inhabitants lost their homes, and virtually the entire business district was destroyed. But it did not take San Francisco long to recover from the disaster, chiefly because the downtown business district was well insured and the city's economy, driven by agriculture, banking, mining, and shipping, rebounded rapidly. By 1910, the city's population stood at 417,000, an *increase* of 22 percent over 1900.[17]

Not to be outdone, smaller industrial cities were flourishing to an extent hard to imagine in light of their subsequent decline later in the twentieth century (and partial revival in some cases). Many cities rose and later fell with the fortunes of a particular industry or company—Detroit (Ford, General Motors), Dayton (National Cash Register, bicycles), Hartford (firearms, Travelers Insurance), Springfield, Massachusetts (firearms, motorcycles, insurance), Pittsburgh (steel), and Rochester (Kodak cameras). Buffalo, an industrial dynamo, was the country's tenth-largest city with over 400,000 inhabitants when it hosted the 1901 Pan-American Exposition—the site of President William McKinley's assassination, followed by Theodore Roosevelt's hasty inauguration. In 2000, Buffalo, no longer a dynamo, ranked fifty-ninth in the nation with 293,000 residents. (A comparison of the top ten cities in 1910 and 2007 appears in table 1.1.) Formerly prosperous manufacturing and retail cities such as Buffalo, Hartford, Providence, Pittsburgh, Cleveland, Baltimore, and St. Louis, despite their subsequent challenges, are still endowed with museums, parks, concert halls, hospitals, and universities donated by wealthy benefactors a century ago.

Even much smaller cities aspired to compete with their peers for population size and civic prestige, albeit with mixed success. The historian Catherine Tumber writes that many turn-of-the-century small industrial cities were unstable in size and function: "Coming into existence with the birth pangs of the American

1.1 Ten Largest U.S. Cities, 1910 and 2007 (by population size)

1910		2007	
New York	4.7 million	New York	8.2 million
Chicago	2.1 million	Los Angeles	3.8 million
Philadelphia	1.5 million	Chicago	2.8 million
St. Louis	687,000	Houston	2.2 million
Boston	670,000	Phoenix	1.5 million
Cleveland	560,000	Philadelphia	1.4 million
Baltimore	558,000	San Antonio	1.3 million
Pittsburgh	533,000	San Diego	1.2 million
Detroit	465,000	Dallas	1.2 million
Buffalo	423,000	San Jose	939,000

Source: 2009 *New York Times Almanac*, 250.

Century, smaller industrial cities were moving targets and hard to define, be-holden to post–Civil War railroad networks in search of market expansion that either induced growth or spelled doom in older towns. As a result, until the 1910s their sizes were unstable, and they were difficult to conceptualize: Were they over-grown towns, or were they underdeveloped cities?"[18]

Smaller cities in strategic locations often flourished. Poughkeepsie, New York, for instance, halfway between New York City and Albany in the Hudson River val-ley, was a thriving regional center of manufacturing and retail shopping in 1900. Despite a modest land area of 5.7 square miles (confusingly surrounded by the much larger Town of Poughkeepsie, a separate governmental unit, where IBM and Vassar College are situated), the city's population grew from 24,000 in 1900, to 35,000 in 1920, and over 40,000 in 1930.[19] (Today it is about 28,000.) Like other small, dynamic cities, Poughkeepsie was united in spirit by civic pride despite a diverse and varied cityscape. And, like those other small cities, Poughkeepsie "sought to enhance its image by promoting the architectural and planning ideals that emerged from the 'White City' designs of the World's Columbian Exposition in Chicago. Public and semi-public architecture of the period sought to advance civic virtues."[20]

The late 1800s and early 1900s saw an explosion of immigrant and working-class urban neighborhoods in cities large and small. New streets of "Victorians," row houses, and apartments sprang up along the expanding arteries of mass transit—subways, trolleys, buses, and elevated trains. Churches, synagogues, temples, most with distinct old world roots, towered above tenements and row houses. As cities expanded, local schools, sewer and water service, parks, and local shopping districts were laid out to serve the new neighborhoods.

The Suburban Exodus Begins

Although "downtowns" thrived in the early 1900s as hubs for offices, hotels, department stores, theaters, museums, clubs, and civic events, the office workers, shoppers, and theatergoers were already beginning to leave center cities in favor of living in the suburbs. Soon after the Civil War, the extension of horse-drawn streetcar lines—soon followed by electric commuter trains in the 1890s—facilitated the beginning of the great suburban exodus. New single-family bungalows and pseudo-Victorian "cottages" clustered near rail stops gradually siphoned white middle-class families from their city apartments and older neighborhoods.

By 1900, Boston's metropolitan area population of one million was about evenly divided between those living in the city proper and those recently relocated to new suburbs.[21] Center cities were still gaining population, but predominantly through immigration from abroad. The number of arriving immigrants reached an all-time high of 1.3 million in 1907 and as of 1910, 13.3 million foreign-born people were living in the United States, comprising one-seventh of the nation's total population.[22] The impulse to move to the suburbs was probably about equally motivated by the *push* of demographic change in older city neighborhoods and the *pull* of new houses available for purchase (instead of rental apartments typical in the city), conveniently connected to downtown by rail. Also contributing to the early suburban migration was the palpable fear of crime, fires, political corruption, saloons, and brothels associated with the older city.

Some new suburbs sought annexation to the parent city to gain access to its public services, as did Roxbury and Dorchester, which joined Boston in the 1870s to gain access to its water system and other public services.[23] But growing antipathy toward the city by the end of the century prompted suburbs to remain independent and provide their own services, or lobby for regionalization of existing systems, such as the metropolitan water, sewer, and parks commissions established by the Massachusetts legislature in the 1890s.[24] But with regionalization of the Boston water system in the 1890s, further annexation ceased; like most northeastern cities, Boston by 1900 was girdled by a ring of suburbs that looked to the city for jobs and entertainment, while preserving their legal autonomy and freedom from the taxes and other burdens of city life.

The private city and suburban development process worked just fine for the wealthy and the rising white middle class of office workers and their families. As Kenneth T. Jackson, Dolores Hayden, and other urban historians have long noted, the suburb enshrined the ideal of gender stereotypes: the male white-collar breadwinner, the female homemaker and mother, and their perfect children. Even the standard house plans for suburban "villas" influenced by Andrew Jackson Downing projected a romantic vision of a pseudo-rural cottage set amid gardens and adjoined by the homes of other white families of similar stage of life and economic status.[25] The barely concealed premise of suburbanization, touted in real estate advertisements of the time, was escape from the poor and the foreign-born, and from the vice, crime, and health threats of the city—for those with the means, and the right skin color and religion (namely, white and Anglo-Saxon Protestant), to leave it all behind.

The "Other Half"

By the early 1900s, the major fault lines of city versus suburb, rich versus poor, "native" versus foreign-born, and WASP versus everyone else were well established. As one looks beyond the showcase parks, skyscrapers, and Beaux-Arts public buildings of downtown, beyond the enclaves of the wealthy and the proliferating suburbs for the white middle class, a very different urban America emerges. In 1900, Manhattan alone contained some 42,700 tenement houses, into which about 1.5 million souls were packed.[26] The Lower East Side was one of the most crowded, impoverished, unhealthy, and dangerous urban districts in the industrialized world. But slum neighborhoods elsewhere in New York and in industrial cities generally were not much better.

Conditions in the festering slums of New York and other late-nineteenth-century cities were sparsely documented or publicized. Foremost among a handful of enterprising journalists who sought to investigate and expose living conditions in tenement neighborhoods was Jacob Riis, a Danish-born New York police reporter turned crusader, who used the new potential of flash photography to depict conditions in the dark back rooms and alleys of the Lower East Side (fig. 1.3). His most influential work, *How the Other Half Lives,* was published initially in *Scribner's Magazine* in 1889 and then expanded as a book in 1890.[27] He memorably described the infamous district "Mulberry Bend" as "the foul core of New York's slums, a vast human pig-sty . . . three acres built over with rotten structures that harbored the very dregs of humanity . . . pierced by a maze of foul alleys," which exposed its inhabitants, especially children, to high rates of mortality from contagious diseases such as tuberculosis that were nurtured by overcrowding and the absence of clean water and sanitation.[28]

Sanitary and health conditions in New York's tenements were more broadly documented by a survey and public exhibit commissioned in 1899 by the housing advocate Lawrence Veiller. The exhibit included detailed maps of poverty and infectious diseases, and more than one thousand photographs. One city block near present-day Chinatown was described in detail as a microcosm of the larger problem. The block contained 39 walk-up tenement houses with 605 apartments housing 2,781 people, 466 of them children less than 5 years of age. Only 264

Fig. 1.3. Jacob Riis, "Bandit's Roost," Mulberry Street, New York City, 1888. Digital Image © The Museum of Modern Art / Licensed by SCALA / Art Resource, NY.

"water-closets" were provided, and hot water was available to only 40 apartments. Of the total 1,588 rooms in this block, 441 had no window, and thus lacked any outdoor light or ventilation. Another 635 rooms opened onto narrow airshafts, rife with noise and the stench of garbage and human waste.[29]

Ironically, many of the tenements thus described were in fact constructed in accordance with a bungled legislative attempt to improve the design of tenement housing. The narrow and fetid air shafts in particular were provided in buildings constructed after the 1879 New Tenement House Law to comply with its prohibition of windowless rooms.[30] However, the "dumbbell" building design favored by the 1879 law—permitting rooms to open onto airshafts rather than streets or courtyards—was itself badly conceived: "The law quickly made the dumbbell tenement a terrible reality for hundreds of thousands of tenants. Landlords packed four families into each of the five or six floors, with each family (plus boarders) squashed into a three-or-four-room apartment, the 'bedrooms' of which measured seven by eight and a half feet. The airshafts—noisy with the quarrels of twenty plus families . . . became garbage dumps and firetraps."[31]

In 1900, the journalist Upton Sinclair disguised himself as an immigrant laborer to investigate the unspeakable conditions in the Chicago meatpacking district, describing the housing surrounding the stockyards as follows:

> Down every street they could see, it was the same—never a hill and never a hollow, but always the same endless vista of ugly and dirty little wooden buildings. Here and there would be a bridge crossing a filthy creek, with hard-baked mud shores and dingy sheds and docks along it; here and there would be a railroad crossing, with a tangle of switches, and locomotives puffing and rattling freight cars rattling by; here and there would be a great factory, a dingy building with innumerable windows in it, and immense volumes of smoke pouring from the chimneys, darkening the air above and making filthy the earth beneath. . . . To the stranger it seemed like a wilderness, a very jungle—a jungle of houses . . . ruled by strange powers.[32]

Early twentieth-century American cities were thus, in Jon Teaford's phrase, "a patchwork quilt rather than a tightly woven fabric."[33] Prairie Avenue in Chicago, lined with the mansions of plutocrats such as meat company owners Armour and Swift, was a short streetcar ride from Sinclair's slaughterhouse "jungle" where their companies' poorly paid, overworked employees lived and died. In New York, Fifth Avenue palaces of the Astors, Vanderbilts, Fricks, and Morgans stood blocks from the fetid, filthy tenements of Harlem and the Lower East Side. Boston's elite Beacon Hill was practically visible from a tenement rooftop in South Boston. Steel mill workers in the Pittsburgh area lived in factory housing in valleys downstream from the hilltop estates and clubs of the company owners and managers.[34] Meanwhile, the daily flow of white male, middle-class commuters between downtown office and suburban garden was barely aware of the poverty and squalor traversed by the streetcar or train line that connected those two foci of the good life.

American Apartheid

In addition to being "ill housed, ill clad, and ill fed" (as Franklin Roosevelt described the foreign-born poor in his second inaugural address), the rural and urban African American population endured the burden of racism and oppression. In 1896, the Supreme Court issued its infamous decision in *Plessy v. Ferguson,* which upheld the doctrine of "separate but equal" treatment of blacks and legalized the Jim Crow segregation of schools, restaurants, trains, public restrooms, and other facilities. The "equality" of such services was fictitious: black waiting rooms, bathrooms, and schools were notoriously inferior to their white counterparts, if they existed at all. These practices prevailed in the South and were common in the North until the landmark 1954 *Brown v. Board of Education* decision overruled *Plessy* regarding school segregation, followed by other rulings and the federal civil rights legislation of the 1960s. (Postwar American apartheid shifted from segregated restrooms to segregated communities with the complicity of federal, state, and local laws and policies, a situation I discuss in chapters 3 and 4.)

One of Theodore Roosevelt's early guests to the White House, five weeks after his inauguration in September 1901, was Booker T. Washington, the distinguished black educator and founder of the Tuskegee Institute—a "world famous figure, revered even by white Southerners."[35] Roosevelt sought Washington's support in nominating moderate blacks to federal civil service positions in both northern and southern states. Reports that a black man dined with the Roosevelts triggered, in Edmund Morris's phrase, "a political hurricane" of hate-filled mail, telegrams, and editorials in leading southern newspapers. Several former Confederate states had already disenfranchised blacks, and lynching was common in both the Old South and border states. Outside Dixie, intensifying racism was provoked in part by a rapid rise of African Americans migrating to northern cities from the Old South. Between 1890 and 1920, the black population of Philadelphia tripled from 39,000 to 134,000, for instance, while the white population grew by only 60 percent.[36] Segregation of travel and restaurants was standard even in the north until midcentury. The world-renowned black contralto Marian Anderson, two years before her epic 1939 Easter performance at the Lincoln Memorial, was denied a hotel room at the Nassau Inn in Princeton, New Jersey. (She stayed overnight instead as a guest of Albert Einstein, an honor not available to other African American travelers.)[37]

The Tragedy of Laissez-Faire

Like Frederick Jackson Turner's western frontier, American cities and suburbs by the early 1900s were bastions of laissez-faire—largely devoid of governmental intervention in the private capitalist economy and the urban communities that it created. On the frontier, this applied to agriculture, grazing, lumbering, mining, and water resources. In the urban context, laissez-faire protected the country's major industries from government regulation regarding public health and safety, consumer protection, workplace conditions, exploitation of labor including women and children, racial discrimination and segregation, price fixing, and hobbling of competition. Many of the nation's largest industries, especially steel, oil, railroads, and banking, were monopolies or oligarchies in the late 1800s, controlled by Gilded Age titans such as John D. Rockefeller, the Astors, the Vanderbilts, Andrew Mellon, Andrew Carnegie, Henry Frick, Jay Gould, J. P. Morgan, E. H. Harriman, and Henry Ford.[38]

As workers began to organize on behalf of better wages, working hours, and workplace conditions, the joint forces of corporate and governmental power opposed the incipient labor movement. The anti-labor tenor of the times was forcefully demonstrated in the deployment of federal troops by President Grover Cleveland to end the 1893 strike of workers against the Pullman Palace Car Company near Chicago—in part to quell a discordant counterpoint to the nearby World's Columbian Exposition then in progress.

Four years earlier, 2,209 people died in the Johnstown, Pennsylvania, flood after a faulty upstream dam broke. The earthen dam and the lake it impounded were the property of a Pittsburgh millionaire's club whose members used the facility as a summer retreat. Despite warnings that the dam was unsafe, no sufficient repairs were made. Those responsible were steel industry owners and executives; the victims were predominantly their employees. The disaster was judicially declared an "Act of God." While the community received an outpouring of contributions from around the country, the dam's owners were exonerated from any liability.[39]

In cities at the turn of the century, freedom from public regulation dominated the layout, construction, and ownership of residential, commercial, and industrial buildings. The profit motive, as reflected in individual and corporate investment decisions, was the overriding determinant of city growth, both horizontally and vertically. A modest exception was the adoption of minimal sanitary regulations and housing standards by some suburban communities. But the incorporation of safe building practices into the middle-class housing market was itself market-driven. It was in the interest of all concerned—lenders, builders, and buyers—that new homes be consistent in appearance and quality of construction in order to maintain the property value of each home and its neighbors. As the urban historian Sam Bass Warner observed: "In the [Boston] suburbs, all the new regulations for plumbing, gas fitting, and building and fire safety were perhaps more important as official affirmations of middle class norms than for policing an occasional offender. With the cooperation and example of the large institutions almost all new suburban building from 1870 to 1900 included safe construction, indoor plumbing, and orderly land arrangement . . . [thus creating] a safe environment for half the metropolitan environment."[40]

Indeed, much of the detailed control of the design, construction, and maintenance of private residential property was governed more stringently by private deed restrictions, voluntarily accepted by homebuyers, than by public land use regulation. When the market demanded uniformity and safety in the construction of buildings, the builders embraced such standards voluntarily. In the absence of such market signals, the anarchy of laissez-faire prevailed.

In the face of the chronic abuses, risks, and occasional catastrophes fostered by laissez-faire, government at all levels was, by definition, impotent. As the historian Roger Biles notes: "City halls and state governments struggled to manage the political, economic, and social affairs of communities steeped in a privatist [*sic*] ethic that favored opportunity over regulation." But they struggled on their own: "In the late nineteenth and early twentieth centuries, neither the state and local government officials responsible for urban governance nor the progressives engaged in various reform activities looked to the federal government for help with city problems."[41]

Two shocking disasters of human origin in the early twentieth century demonstrated the terrible costs of laissez-faire in the industrial age. The March 25, 1911, a fire at the Triangle Shirtwaist Factory in New York killed 146 young women who were trapped in a ten-story sweatshop behind doors kept locked by the management. Primitive movie footage recorded and publicized images of victims leaping to their death from upper stories. The sinking of the *Titanic* a year later on April 15, 1912, further inflamed public outrage against corporate hubris: over 1,600 people died due to negligent design and operation of the vessel by its British owners, and a woefully inadequate supply of lifeboats. These catastrophes each served to arouse the national media and the educated public to the perils of laissez-faire and laid a foundation for the gradual enlargement of the public role in curtailing the excesses of private capitalism and urban development in the United States over the coming century.

Precursors to Reform

The twentieth century, of course, did not start with a blank slate in terms of public measures to temper the effects of rapid urban growth and industrialization. The path-breaking revelations by Jacob Riis, Lawrence Veiller, and others helped to stimulate support for public regulation of housing. As such, they contributed to the first of three major strands of public and private response to urban squalor: (1) regulation, (2) redevelopment, and (3) relocation.[42]

Regulation concerned housing and building practices, fire safety, and other urban "housekeeping" issues—each a cautious departure from the prevailing faith in laissez-faire. While efforts to regulate aspects of the urban environment date back to ancient Greece and Rome, as well as medieval Europe (although they were often ignored), the genesis of modern building regulations has been attributed to the 1667 Act for Rebuilding London passed within months after the Great London Fire of September 2–7, 1666. The act specified fire-resistant building practices, as recommended by a royal commission appointed after the fire.[43] In the 1830s, the British sanitary reformer Edwin Chadwick and his medical colleagues investigated conditions in the slums of London, Manchester, and other cities. Their reports, and the general fear of infectious disease spreading from tenements to "respectable" districts, prompted Parliament to adopt the Public Health Act of 1848. In New York City, investigations by John Griscom, inspired by Chadwick's work in England, eventually led to the adoption of the 1866 Metropolitan Health Act,[44] followed by the New Tenements Act of 1879 and other (equally timid) laws adopted in other cities.

Redevelopment referred to what we would now call "urban infrastructure." During the nineteenth century, examples of redevelopment included:

- Urban parks inspired by Frederick Law Olmsted and Calvert Vaux's Central Park in New York and the chain of greenspaces in Boston known as the Emerald Necklace designed by the Olmsted firm.
- Development of external public water sources such as New York's Croton River project and Boston's Lake Cochituate system, both designed by John Jervis in the 1840s and later much expanded.
- Engineering projects to convey urban sewage away from cities and their water sources, such as the reversal of the Chicago River.
- Transportation improvements, including paved and lighted streets, bridges, railroads, subways, and elevated trains.
- The founding of civic institutions (museums, concert halls, hospitals, universities, and libraries) established and endowed primarily by Gilded Age philanthropists such as Andrew Carnegie, John D. Rockefeller (Sr. and Jr.), the Vanderbilts, the Mellon brothers, and their counterparts in smaller industrial cities.

Relocation involved experiments in Europe and the United States with the creation of small planned industrial communities outside of large cities to which a company's factory and workforce could be relocated. The best-known American example is Pullman, Illinois, planned and constructed at the behest of the Chicago railroad tycoon George Pullman to provide decent housing and a "moral" environment for employees of his sleeping car company. His paternalism backfired, however: tired of low wages, high prices at the company food store, and no source of alcoholic beverages, the workers went on strike in 1893. After federal troops put down the strike, the town eventually withered into a quaint relic of Victorian idealism gone awry.

With these and other antecedents to build on, the Progressive movement confronted a broad spectrum of perceived abuses, including corporate monopolies ("trusts"); political corruption and big city "machines"; free-wheeling financial practices that led to the Panics of 1893 and 1907, and eventually the 1929 stock market crash; exploitation of labor; moral decay and vice; and risks to public health and safety. The outgrowths of efforts to address these and other social problems yielded a rich but sometimes conflicting array of social movements. Although they differed widely in their goals, constituency, vocabulary, and style, each shared an implicit faith in the power of a modern society to formulate and pursue necessary actions to improve the quality of life for its citizens.

Chapter 2

Competing Visions in the Progressive Era

Make no little plans. They have no magic to stir men's blood and probably themselves will not be realized. Make big plans; aim high in hope and work.
—Daniel H. Burnham, 1907

It is so easy for the good and powerful to think that they can rise . . . by pursuing their own ideals, leaving those ideals unconnected with the consent of their fellow-men. —Jane Addams, 1912

The assassination of President William McKinley followed by the inauguration of Theodore Roosevelt in Buffalo in 1901 would mark an ideological watershed in the evolution of American cities, along with the nation's economy and political system. In contrast to McKinley—a Civil War veteran and traditional pro-business Republican from Canton, Ohio—the young Roosevelt would be the archetypal "Progressive Republican" (an extinct species in today's political climate). Born in New York City, educated at Harvard and Columbia, a self-anointed hero of the Spanish-American War, a former everything—New York City police commissioner, U.S. Civil Service commissioner, assistant secretary of the navy, governor of New York, vice president—and a prodigious writer and adventurer, "TR" embodied the social and economic ideals of the Progressive movement which energized the metropolitan intelligentsia of his time. According to the movement's leading historian, Richard Hofstadter, progressivism reflected the "broader impulse toward criticism and change that was everywhere so conspicuous after 1900. . . . [I]t was a rather widespread and remarkably good-natured effort of the great part of society to achieve some not-very-clearly-specified self-reformation. Its general theme was the effort to restore a type of economic individualism and political democracy that was widely believed to have existed earlier in America and to have been destroyed by the great corporation and the corrupt political machine."[1]

No president at that time, even with the reformist zeal of Theodore Roosevelt, had much direct influence over cities, which were legally created and empowered by state legislatures and largely left to their own often corrupt devices. However, a new generation of educated, wealthy, and well-traveled members of the nation's upper classes believed change for the better was within the grasp of cities and their citizenry. In the early 1900s, two such initiatives—loosely sharing the banner of "progressivism" with other reform efforts of the time—crystallized around two parallel crusades: the patrician and aesthetics-dominated City Beautiful movement and the strongly feminist and poverty-oriented settlement house movement.

Although both emanated from Chicago's South Side in the 1890s, two more contrasting social movements could scarcely be imagined. City Beautiful emphasized the role of public art and architecture, chiefly in the neoclassical style, as a medium to uplift the aesthetic, economic, and moral well-being of cities. It strongly focused on the city center and was dominated by men of substance (financial and corporeal) epitomized by the celebrated architect Daniel H. Burnham. The settlement house crusade, led by the indefatigable Jane Addams and her cadre of educated women colleagues, struggled to address social ills and injustice. Thus were counterposed the competing paradigms of urban improvement which dominated America over most of the twentieth century and linger on today.

The City Beautiful Movement

The City Beautiful era famously began with the 1893 Chicago World's Columbian Exposition. The story of the fair's creation has been often recounted, most colorfully in Erik Larson's *The Devil in the White City*.[2] At the behest of the city leadership who wanted to showcase Chicago's recovery from its Great Fire of 1871, the project was created by the nation's leading architects under the direction of Daniel H. Burnham, with the landscape design by Frederick Law Olmsted. During its seven months' duration, the fair hosted some twenty-seven million visitors to an exuberant but temporary "White City" of some two hundred glistening neoclassical exhibit buildings, ornamental grounds, reflecting pools, a lagoon with Venetian gondolas, a Court of Honor ringed with statuary, the frivolities and excitement of the Midway, and the world's first Ferris wheel (with a yet-to-be surpassed capacity for carrying more than 2,000 people at a time)—all powered and illuminated by the new technical wonder of electricity (fig. 2.1).

The fair left bedazzled visitors with two take-home messages: first, that cities of the future should be designed by experts to be aesthetic, orderly, and functional (approximately in that order); and second, that the favored architectural style for aspiring cities, mimicking Paris, Vienna, and other European capitals, was neoclassicism. The high priests of this City Beautiful religion—who included William Morris Hunt, H. H. Richardson, Stanford White, and Daniel H. Burnham—used the 1893 Chicago Columbian Exposition to showcase their training in neoclassicism at the École de Beaux-Arts in Paris. Georges Haussmann's midcentury transformation of Paris was the aesthetic and functional model for Chicago's "White City" and its imitators across America. Downtown city plazas from this era were relentlessly geometric and focused on statues or fountains, surrounded by formal gardens and paved pedestrian spaces. From Washington, DC, to Cleveland to Denver to Seattle, the nation's older city centers still are dominated

Fig. 2.1. The Court of Honor at the center of the 1893 Columbian Exposition. Courtesy of the Chicago History Museum. Neg. ICH1-02526; C. D. Arnold, photographer.

by public spaces and pompous architecture from the City Beautiful era: ancient Egypt, Greece, and Rome meet Main Street.

The oracle of the City Beautiful movement was Charles Mulford Robinson, the author of two influential treatises: *The Improvement of Towns and Cities; or, the Practical Basis of Civic Aesthetics* in 1901 and *Modern Civic Art: or The City Made Beautiful* in 1903. How could any cultured person of the day resist such Wagnerian odes to the City Beautiful as the following, from *Modern Civic Art*:

> There is the promise of the dawn of a new day. The darkness rolls away, and the buildings that had been shadows stand forth distinctly in the gray air. The tall facades glow as the sun rises; their windows shine as topaz; their pennants of steam, tugging flutteringly from high chimneys are changed to silvery plumes. Whatever was dingy, coarse, and ugly, is either transformed or hidden in shadow. The streets bathed in the fresh morning light, fairly sparkle, their pavements from upper windows appearing smooth and clean.

There seems to be a new city for the work of a new day. . . . As when the heavens rolled away and St. John beheld the New Jerusalem, so a vision of a new London, a new Washington, Chicago, or New York breaks with the morning sunshine upon the degradation, discomfort, and baseness of modern city life.[3]

The Improvement of Towns and Cities more technically defined how each element of a city—land and water approaches, administrative center, business district, streets, parks, residential areas, and tenements—should be artistically improved to achieve a "New Jerusalem" on the Hudson, Lake Michigan, or wherever. As taught to this generation of architects at the École de Beaux-Arts in Paris, the architectural vocabulary is relentlessly neoclassical: domes, arches, columns, porticoes, obelisks, sculpture, heroic statuary, fountains, cupids, nymphs, reflecting ponds, formal gardens, trimmed hedges, and manicured lawns.

But behind the frothy prose and illustrations, Robinson and other City Beautiful polemicists offered more than simply homage to European neoclassicism. According to the movement's leading historian, William H. Wilson, Robinson's books "revealed the rich texture of the City Beautiful movement, . . . [and] dealt with virtually all aesthetic and practical urban developments, historic and recent, except for subsurface drainage and transportation. Urban sites, watercourses, playgrounds, street patterns, paving, lighting, and sanitation . . . passed under review. He addressed the need for controlling urban smoke, noise, and billboards."[4]

The movement gained momentum with the 1902 Senate Parks Commission (known as the McMillan Commission) plan for the nation's capital.[5] Prepared at the behest of Congress as the overseer of the District of Columbia, the commission retained the same arbiters of urban design who had created the 1893 fair: architects Daniel H. Burnham and Charles F. McKim, landscape architect Frederick Law Olmsted Jr. (representing his ailing father), and sculptor Augustus Saint-Gaudens. Enthusiastic support by the new American Institute of Architects at its Washington, DC, convention in 1900 reflected a triumph for art and architecture proponents of the McMillan Commission over the more utilitarian plans for the District developed by the Army Corps of Engineers.[6] In the afterglow of victory in the Spanish-American War and the inauguration of President Theodore Roosevelt, the McMillan Plan projected a vision of Washington as a world imperial capital. Under its guidance, the National Mall with its monuments and museums would become the nation's foremost public space, hallowed by celebrations, protests, and huge gatherings at times of national joy and sadness. The Lincoln Memorial, a late City Beautiful–era Greek temple, would be the striking venue for two civil rights milestones: the nationally-broadcast recital by Metropolitan Opera mezzo-soprano Marian Anderson on Easter Sunday 1939 and Martin Luther King Jr.'s "I Have a Dream" address on August 28, 1963.

The 1909 *Plan of Chicago,* one of the seminal documents in American planning history, represents both the apotheosis and demise of the City Beautiful movement as a force behind progressive urban improvement efforts. Daniel H. Burnham and Edward H. Bennett prepared the plan for the Commercial Club of Chicago, which was scarcely a radical or socialist organization. A photograph of the principals involved in the plan's preparation depicts fifteen white, affluent, portly, well-attired gentlemen wining and dining in Gilded Age splendor (fig. 2.2). According to the historian Carl Smith, of the thirty-two club members chiefly responsible for the plan, "virtually all were Protestant and Republican, and every one was white and male [and all of them] believed in the ethos of the City Beautiful Movement."[7] The plan epitomized the symbiosis of capitalism and "efficiency" that underlay the mainstream progressive urban agenda.

Urban beautification, of course was a central objective of the *Plan of Chicago*. Burnham brilliantly argued that "beauty" may be understood not only in narrow aesthetic terms, but also in the broader sense of a city that is "well-ordered and convenient." Moreover, such order and convenience not only anchors those of "means and taste," but draws in others: "The very beauty that attracts him who

Fig. 2.2. Commercial Club patricians overseeing the preparation of the 1909 *Plan of Chicago*. Courtesy of the Chicago History Museum. Neg. ICH1-03560; photographer unknown.

has money makes pleasant the life of those among whom he lives, while anchoring him and his wealth to the city. The prosperity aimed at is for all Chicago."[8] Having thus stated what his corporate sponsors expected (anticipating the "trickle down economics" of the Reagan and Bush administrations), Burnham shifts from City Beautiful dogma to broader progressive themes more in sympathy with Jane Addams and Hull House a few blocks away:

> Thoughtful people are appalled at the results of progress; at the waste in time, the toll of lives taken by disease when sanitary precautions are neglected, and at the frequent outbreaks against law and order which result from narrow and pleasureless lives. So that while the keynote of the nineteenth century was expansion, we of the twentieth century find that our dominant idea is conservation. . . .The constant struggle of civilization is to know and to attain the highest good; and the city which brings about the best conditions of life becomes the most prosperous.[9]

In its actual content, the *Plan of Chicago* was far more sweeping, both geographically and functionally, than the Senate Parks Commission Plan for Washington of seven years earlier. The lavishly illustrated volume (clearly intended for the mahogany coffee tables of Chicago's Prairie Avenue and Gold Coast set) marked a transition from the "pure" City Beautiful to a more pragmatic "City Functional" paradigm of city improvement. The *Plan*'s core and quintessential City Beautiful element—a hulking forty-story Beaux-Arts civic center as the apex of radiating boulevards—was fortunately never built (fig. 2.3). (The civic center in fact would have destroyed the Halsted neighborhood served by Hull House and likely the Hull House complex itself[10]—a task accomplished in the 1960s with urban renewal and highway construction.)

Nonetheless, the city would be reshaped by realizing many of the *Plan*'s more pragmatic elements, including: a twenty-three-mile chain of lakefront parks, lagoons, and public facilities; over sixty thousand acres of parks, parkways, and forest preserves; new bridges, boulevards, and widened arterial streets; a program to clean up and straighten the course of the Chicago River through downtown; the consolidation of freight and passenger rail systems and mass transit; and the expansion of Grant Park as a hub of cultural and educational institutions.[11]

Major aspects of the *Plan of Chicago* were popularized in Chicago public schools with an eighth-grade-level civics textbook created by Walter Moody, a member of the 328-member commission appointed in 1909 to implement the *Plan*. The book, called *Wacker's Manual for the Plan of Chicago*, influenced a generation of city taxpayers to support such projects as the completion of the city's lakefront park system; construction of the double-decked Wacker Drive along the Chicago River and the 12th Street Bridge across the rail yards south of the central business district; the improvement of railroad interconnections; and, at a regional scale, the establishment of the Cook County forest preserve system in 1914.[12] Although directed by one of the high priests of the City Beautiful, the *Plan of Chicago* effectively

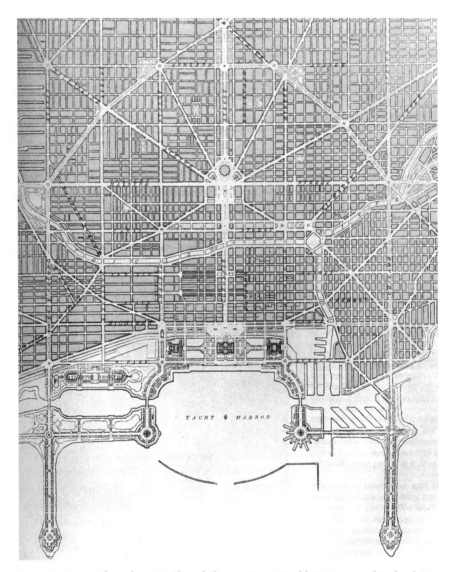

Fig. 2.3. Excerpt from the 1909 *Plan of Chicago.* From Harold M. Mayer and Richard C. Wade, *Chicago: Growth of a Metropolis* (Chicago: University of Chicago Press, 1969), 277.

marked the end of purely aesthetic urban design and a transition toward modern technical and pragmatic city planning.

Both the aesthetic and pragmatic aspects of the City Beautiful would affect a broad swathe of American cities through the many civic designs of the landscape architect and planner John Nolen, the author of *Replanning Small Cities:*

Six Typical Studies (1912) and *New Towns for Old* (1927). According to Catherine Tumber, Nolen "undertook almost 400 commissions before his death in 1937, concentrating on planning smaller northern cities, from Reading, Pennsylvania, and Montclair, New Jersey, to Madison, Wisconsin, and Bridgeport, Connecticut, before turning his attention to projects in the South and West."[13]

Neoclassicism persisted into the New Deal, chiefly in the work of John Russell Pope, the designer of the National Archives building, the Jefferson Memorial, and the National Gallery of Art. The conservatism of this style clashed with the fast-spreading modernist style characterized by dramatic geometric shapes, the absence of decoration, and the use of steel and glass. The series of world's fairs held in the United States during the depression, discussed later in this chapter, offered a mix of "slimmed classicism" ("Mussolini-style") and true modernism inspired by the works of Frank Lloyd Wright, Le Corbusier, and Mies van de Rohe.[14]

For better or worse, innumerable legacies of the City Beautiful era survive today. Some are treasured as incomparable public spaces and buildings, such as the National Mall and Union Station in Washington, DC, Chicago's Grant Park with its dazzling Buckingham Fountain, and New York's Carnegie Hall, Public Library, and narrowly rescued Grand Central Terminal. Less revered—and sometimes reviled—are innumerable state capitols, city halls, courthouses, train stations, government offices, and university buildings which survive as incongruous ornaments or white elephants amid the clutter of later growth and redevelopment. The City Hall and Symphony Hall in Springfield, Massachusetts, for instance, occupy twin colonnaded Greek temples separated by an Italian campanile, now adjoined by an elevated interstate highway and a defunct downtown mall.

The mixed messages of the City Beautiful—aesthetics and functionality—would conflict repeatedly over time as grandiosity often trumped practicality. This tension would dominate urban design throughout the twentieth century long after neoclassicism lost its appeal in favor of other architectures of power, such as Modernism and its offshoot, the International Style, as popularized after World War II in the work of Ludwig Mies van der Rohe, Walter Gropius, Philip Johnson, and others.

The Settlement House Movement

As yang to the City Beautiful yin, the settlement house movement in the United States originated with the founding of Hull House on the Near West Side of Chicago by Jane Addams (fig. 2.4) and Ellen Gates Starr in 1889. The settlement house, like sanitary reform, was inspired by an English precedent: Toynbee House in London. After visiting that facility during a European Grand Tour, The college-educated and independently wealthy Addams purchased a pre–Civil War mansion that had become surrounded by tenements near the Chicago stockyards.[15]

Fig. 2.4. Jane Addams, ca. 1910. Photograph JAMC_0000_0004_2792, Jane Addams Memorial Collection, Special Collections & University Archives, University of Illinois at Chicago Library.

In *Twenty Years at Hull-House* she described the abysmal squalor of this six-mile-long district largely inhabited by immigrants in 1910: "The streets are inexpressibly dirty, the number of schools inadequate, the street lighting bad, the paving miserable and . . . and the stables foul beyond description. Hundreds of houses are unconnected with the street sewer. . . . Many houses have no water supply save the faucet in the back yard, there are no fire escapes, the garbage and ashes are placed in wooden boxes which fastened to the street platforms. . . . The houses of the ward, for the most part wooden, were originally built for one family and are now occupied by several. . . . [M]any houses have no water supply save the faucet in the back yard, there are no fire escapes, the garbage and ashes are placed in wooden boxes which are fastened to the street pavements.[16] As Jacob Riis's *The Other Half* focused attention on New York's Lower East Side, the Stockyard district soon became an object of national outrage, both through the growing fame of Hull House itself and through the publication of Upton Sinclair's *The Jungle* in 1906—a book that Theodore Roosevelt personally pressed on his White House visitors and friends.[17]

The Hull House charter of 1889 defined a threefold mission for the program: "To provide a center for a higher civic and social life; to institute and maintain educational and philanthropic enterprises, and to investigate and improve the conditions in the industrial districts of Chicago."[18] The model for Hull House, emulated by dozens of similar facilities elsewhere, was to recruit educated and idealistic young women from the "comfortable classes"—like Addams and Gates themselves—who would agree to live at the settlement house for substantial periods of time as volunteers to work with and be mentors to the immigrants living in the surrounding neighborhoods. By cultivating the wives of wealthy businessmen (such as the members of the Commercial Club), Addams and Gates quickly attracted substantial support and by 1911 were able to expand the project to a campus of thirteen buildings, including in addition to Hull House itself, a gymnasium, coffee house, library, auditorium, art gallery, cooperative apartments, meeting spaces, and classrooms. According to Addams's biographer Louise Knight: "Famous visitors flocked to the house. . . . The crusading journalist Ida Tarbell, [nemesis of John D. Rockefeller] recalled that Hull House was the place everyone visiting Chicago from overseas wanted to see."[19] In 1914, Walter Lippmann described Hull House as "friendly . . . beautiful, sociable and open" and the embodiment of what "civilization might be like."[20]

Hull House early established the city's first kindergarten and hosted countless social clubs and classes in English, cooking, sewing, carpentry, and other life skills for the neighborhoods youths and adults. It sponsored programs of arts, theater, poetry, and literature, drawing on the broader cultural resources of the city and the nearby University of Chicago (founded in 1892). Hull House created the city's first playground in 1893, and Addams would head the Playground Association of America founded in 1906.[21]

As a product of Jane Addams's powerful intellect and spirit, Hull House was much more than a safe, warm, and welcoming refuge from the rigors of tenement life for the newly arrived immigrants. Profoundly influenced by the writings and philosophy of Leo Tolstoy, whom she visited in Russia in 1895, Addams eschewed "benevolence," which she viewed as "a selfish, arrogant, and antidemocratic ethic that needed to be retired from contemporary life."[22] Rather, she embraced the Tolstoyan conviction that "good works" such as Hull House were "a mere pretense and travesty of the simple impulse 'to live with the poor' as long as the residents did not share in the common lot of hard labor and scant fare."[23] Today, we might call that "walking the walk; not just talking the talk."

From her Hull House home and base where she lived until her death in 1935, Jane Addams walked many walks. She early championed labor reform, supporting the workers in the 1894 Pullman strike, and thereafter lending her influential voice and pen to the crusade for state and federal labor legislation. She mobilized her young associates to conduct field investigations of housing and sanitary conditions in the surrounding neighborhoods, leading to the publication of *Hull House Maps and Papers* in 1893 and many subsequent studies. These were not academic exercises: one of their disclosures was the prevalence of uncollected garbage, horse carcasses, and human waste in the district's streets and alleys. Addams shamed the mayor into appointing her as garbage inspector for the Nineteenth Ward at a salary of $1,000 per year in place of the political hack who had previously held the position, one that she discharged diligently by her own account.[24]

During the two decades between the founding of Hull House and the 1909 *Plan of Chicago,* the city doubled in population from one to two million,[25] with the majority of newcomers being low-income migrants from Europe and rural America. While both Jane Addams and Daniel Burnham shared a progressive-minded zeal to "improve" the city of Chicago, their perspectives priorities could scarcely have been more different. Carl Smith observes that, "Jane Addams and other settlement leaders chose to live in the neighborhoods and improve them from within, in collaboration with people already living there. Burnham [who lived in the leafy suburb of Evanston] and the Commercial Club sought to impose 'order' from afar, represented by bird's-eye views in the *Plan of Chicago,* beautiful but dehumanized representations of order, requiring wholesale demolition of existing neighborhoods."[26]

Thus were starkly juxtaposed the two worldviews of urban improvement that would clash throughout the coming century: the top-down, expert-driven imposition of a uniform city vision congenial to the urban elite (the City Beautiful and its later analogues such as urban renewal)[27] and bottom-up, pragmatic, and democratic efforts to solve problems at the block and neighborhood level (the settlement house movement and its descendents such as the Local Initiatives Support Corporation, Habitat for Humanity, and urban agriculture movements of recent decades). Partisans of the latter worldview, in its many manifestations, would wage the "struggle for people, place, and nature" in decades to come.

The implications of these competing perspectives for Chicago in its first decade are illuminated in Janice Metzger's intriguing 2009 book, *What Would Jane Say?* Jane Addams was initially consulted in 1907 by Burnham on widening Halsted Street past Hull House—an idea she did not oppose provided it was "part of a well-considered scheme for the general improvement of the city."[28] Addams's opinions on the proposals were not solicited aside from on that relatively local matter, and, in fact, the final *Plan* proposed to located its grandiose Civic Center virtually on top of the by-then world famous Hull House and its neighborhood (see fig 2.2 above).[29] Drawing on the copious writings of Jane Addams and her colleagues, Metzger speculates on how the women of Hull House would have responded had they been invited to comment on the *Plan* and how it addressed such issues as parks, transportation, immigrants and labor, public health, housing and neighborhoods, education, and the juvenile justice system.[30] She reasonably infers that they would have viewed much of the *Plan* as serving the interests of the wealthy and powerful, while largely ignoring the needs of the city's neighborhoods and the poor.[31] On the issue of parks, for instance, the settlement house women would likely have opposed grandiose downtown parks that would burnish the city's self-image without equivalent provisions for neighborhood parks and play-grounds to serve ordinary families who could not afford the cost of a roundtrip tram ride to visit a showplace park downtown.[32]

With Hull House on a solid footing, Jane Addams assumed a much broader role as a widely published and admired public intellectual. Not surprisingly, she played a strong role in the women's suffrage movement, which gradually achieved its goal state-by-state, culminating in adoption of the Nineteenth Amendment to the U.S. Constitution in 1920. She was an outspoken pacifist, attracting much public scorn for her opposition to intervening in World War I. Her role as President of the International League for Peace and Freedom and related activities earned her the 1931 Nobel Peace Prize (as the first woman Nobel laureate). She also was an early and powerful voice for civil rights, a role that she symbolically passed on shortly before her death in 1935 to Eleanor Roosevelt at a Washington, DC, banquet.[33]

The settlement house movement spread rapidly from its base in Chicago. By 1900, there were more than one hundred such institutions in the United States, a number that doubled by 1905 and doubled again to four hundred by 1910.[34] Addams proselytized many fellow women leaders to embrace the Hull House strategy, notably Florence Kelley, Julia Lathrop, and Lillian Wald, the founder of New York City's Henry Street Settlement. Frances Perkins, a Hull House partici-pant and labor reform specialist, became the first woman cabinet member as FDR's secretary of labor. Catherine Bauer (in later years known by her married name, Wurster) and Edith Wood reflected the spirit of the settlement house movement in their advocacy for public housing, which led to the Housing Act of 1937.[35]

The Jane Addams Hull House Association today is the largest social service agency in Chicago. It honors Addams's work and vision through dozens of social programs housed in three neighborhood locations. Hull House itself survives with one other building as a National Historic Landmark and museum on the campus of the University of Illinois at Chicago.

The City Planning and Zoning Movement

In 1909, the year the *Plan of Chicago* was released, another watershed event signaled the decline of urban beautification as the dominant paradigm of city improvement efforts. The First National Conference on City Planning and Congestion held in Washington, DC, that year attracted a broad range of progressive activists con-cerned only incidentally with aesthetics, and more directly with the plight of im-migrants, the working poor, and the habitability of cities. During the first decade of the century, an amalgam of progressives of various stripes coalesced around "anti-congestion," a catch-all term that included overcrowding of tenements, depriva-tion of sun and daylight, of fresh air and recreation, excessive building density, crime, fires, traffic delays and accidents, and in the largest cities, the proliferation of downtown skyscrapers. The "anti-congestion" agenda owed much to settlement house investigations of slum conditions, and to Lawrence Veiller's 1900 tenement house study and exhibit on the spatial relationship of overcrowding, infectious disease, and poverty. Another influence was the playground movement, which originated with the establishment of model playgrounds at Hull House in Chicago and the Henry Street Settlement in New York and led to the formation in 1906 of National Playground Association of America. The "anti-congestion" agenda also was shaped by the reports of crusaders such as Jacob Riis, Ida Tarbell, Lincoln Steffens, and Frederick Howe.

The 1909 conference was organized by a young social activist, Benjamin C. Marsh, the executive secretary of the fledgling Committee on Congestion of Population based in New York City. Marsh had been influenced by Frederick Howe's *The City: The Hope of Democracy* (1905), which extolled the new practices of city planning in Germany. After collecting his own impressions from European cities, particularly Frankfurt, Marsh published *An Introduction to City Planning* urging adoption of the German experience in imposing public control over urban planning and the design of streets, neighborhoods, and buildings.[36] Unlike the *Plan of Chicago*, which largely called for public improvements, Marsh advocated public oversight of private land development through tax measures, municipal acquisition, and regulation of height and bulk of new buildings.

The 1909 conference was a "Who's Who" of innovative thinkers on urban problems of the time (other than Daniel Burnham who did not attend): "The forty-three conferees met in an air of excitement and hope. Many of the nation's leaders in urban affairs attended, including Frederick Howe, Jane Addams, . . . John Nolen, [and] Frederick Law Olmsted, Jr. . . . Representatives of municipal art, social work, architectural, civil engineering, and conservationist groups also attended. The meeting vividly reflected the many interest groups concerned with city planning at the time."[37] In Burnham's absence, the landscape architect Robert Anderson Pope railed against the orthodoxy of the City Beautiful: "We have rushed to plan showy civic centers of gigantic cost . . . brought about by civic vanity, . . . when pressing hard-by, we see the almost unbelievable congestion with its hideous brood of evil, filth, disease, degeneracy, pauperism, and crime. What external adornment can make truly beautiful such a city?"[38] Instead of "showy civic centers," Pope urged decentralizing new development and more equitable distribution of land values; widening of streets and establishment of radial and belt thoroughfares; and the adoption of land use zoning as practiced in Germany to regulate "building heights, depth of blocks, number of houses per acre, and land speculation with all its attendant evils." Pope echoed a remarkable belief among urban reformers at this time that city planning was essential to national and ethnic survival.[39]

The fledgling planning movement steadily gained momentum after the 1909 conference. A follow-up National Conference on City Planning was held in Rochester,

New York, in 1910 (with the term "congestion" conspicuously dropped from its title). As the movement spread intellectually and geographically, Benjamin Marsh's focus on "congestion" lost salience for being too New York City–centric. Furthermore, Marsh's standing among business and financial leaders was undermined by his passionate advocacy of Henry George's concept of a "single tax on land."[40] Under new leadership, the 1910 conference evolved into an ongoing organization of the same name, one ancestor of today's American Planning Association. In the process, the planning movement became less ideological and more "professional," that is, concerned with "data, statistics, techniques, management, standards, efficiency, and evaluation."[41]

In the urban planning context, the Holy Grail of progressive lawyers and planners was to institute regulation of the use of private property by local public authorities ("land use zoning"), as practiced in Germany.[42] Land use regulation involved public limitation of a property owner's freedom to use land and build on the land as he or she wished, without compensation for any resulting loss of value. It was assumed, realistically, that reimbursing property owners for loss of value would be far too expensive to the government and its taxpayers. The major hurdle facing the advocates of land use zoning, therefore, was eventually to persuade the U.S. Supreme Court to declare constitutional the public authority to regulate private land use.

The primary legal basis in support of zoning was the "police power," the inherent power of the sovereign to control private behavior to the extent necessary to protect the public health, safety, and welfare. While no one would deny the power of government to prohibit murder or arson, would that power extend to preventing construction of a saloon in a residential neighborhood or requiring a minimum lot size to build a home?

Countering the police power argument, property rights advocates (then as now) invoked the Fifth Amendment, which states in part: "nor shall private property be taken for public use without just compensation." Those twelve words reflected the framers' fear of sovereign preemption of private property rights, as expressed most forcefully by their legal mentor, the distinguished English jurist Sir William Blackstone in 1768: "There is nothing which so generally strikes the imagination, and engages the affections of mankind as the right of property; or that sole and despotic dominion which one man claims and exercises over the external things of the world, in total exclusion of the right of any other individual in the Universe."[43] This hyperbolic statement, if taken literally, would clearly nullify public land use regulations as curtailing the owner's "sole and despotic dominion" without compensation. But the Anglo-American legal tradition has proven ingenious in adapting old doctrines to changing needs and conditions of society. The acceptance of land use zoning required such a paradigm shift by the U.S. Supreme Court. Would the eighteenth-century fear of an oppressive monarch defeat a twentieth-century progressive strategy for bringing order to city development?

The issue was first articulated in the epic 1922 Supreme Court decision in *Pennsylvania Coal Co. v. Mahon*,[44] involving a new Pennsylvania law that limited the right of coal producers to mine beneath inhabited areas. The statute banned "pulling of pillars" supporting the surface above coal mines, a practice that was causing collapse of the surface and endangering the lives of miners. The plaintiff coal company challenged the law as "taking" the value of coal required to be left in the ground without compensation by the state. In an 8–1 decision recited ad nauseum by land use lawyers ever since, Oliver Wendell Holmes Jr. acknowledged that public regulation of private property without compensation is necessary and constitutional under certain circumstances: "*The general rule at least is, that while property may be regulated to a certain extent, if regulation goes too far, it will be recognized as a taking.*"[45] This rule in effect required a balancing of public and private impacts to determine whether or not a regulation is "reasonable" (i.e., constitutional). Although the decision upheld the coal company's "takings" claim, it was not a reversion to Gilded Age laissez-faire capitalism. Holmes's balancing test—that public regulations are constitutional if they don't "go too far"— was a long step beyond public-be-damned conservatism. In actuality, it opened the door for governments to use regulation if applied "reasonably," for example, if based on careful planning studies. Thus was born the modern practice of land use regulation—a hornets' nest of legal disputes and the source of full employment for land use lawyers—but a key component of the progressive city planning agenda.

Before 1920, Chicago and New York remained the epicenters of intellectual leadership and experimentation with land use regulation. But New York's early embrace of zoning was unrelated to living conditions of tenement districts and the "other half." Instead, wealthy Fifth Avenue merchants embraced zoning as a means of protecting their trade and property values from intrusion of multistory garment factories and offices (such as the ten-story Triangle Shirtwaist sweatshop that burned in 1911).[46] Height limits and tapering of skyscrapers was proposed to reduce overshadowing and overcrowding in business districts. Another issue was the helter-skelter nature of residential development in the outer boroughs newly connected by subway and elevated lines to downtown Manhattan.

These and other concerns coalesced in the nation's first land use zoning law, which was adopted with startling alacrity by the city in 1916 and quickly elsewhere throughout the country. The progressive argument for zoning presupposed that it would be exercised primarily by large cities within which all needed land uses could be accommodated somewhere, and that it would be applied "in accordance with a comprehensive plan." Such locally prepared "master plans" would address the city's broad problems and recommend solutions to be implemented in part through zoning laws. No one yet knew in the early 1920s, however, whether the wildly popular practice of land use zoning was "reasonable" and constitutional according to the Holmes balancing test.

Ironically, the landmark Supreme Court decision on zoning in 1926 involved not an expert-advised central city, but a scruffy industrial Cleveland suburb: Euclid, Ohio. A prescient trial court decision by the U.S. District Court for the Northern District Ohio in 1924 held Euclid's fledgling zoning law to be *unconstitutional* as an attempt to "to classify the population and segregate them according to their income or situation in life."[47] If the U.S. Supreme Court had affirmed this decision on appeal, the face of metropolitan America might look quite different. However, the high court reversed it based on arguments from national planning organizations that rushed to defend their new and widespread invention as necessary and reasonable to bring order to fast-growing cities. In a 6–3 decision upholding the constitutionality of zoning, the majority memorably voiced the progressive view of the role of law in a complex society: "While the meaning of constitutional guaranties never varies, the scope of their application must expand or contract to meet the new and different conditions which are constantly coming within the field of their operation. In a changing world, it is impossible that it should be otherwise."[48]

The Supreme Court not only upheld land use zoning, it also armed zoning authorities with a presumption in their favor in future challenges by property owners: "If the validity of the legislative classification for zoning purposes be fairly debatable, the legislative judgment must be allowed to control."[49] In other words, local zoning authorities have the benefit of the doubt; challengers (such as civil rights groups in the 1970s) must overcome that presumption with proof that a zoning regulation is so arbitrary and discriminatory that it is not even "fairly debatable."

Euclid was essentially the last word on zoning from the Supreme Court for six decades. Understandably, the court declined to review further zoning cases in "tedious and minute detail,"[50] leaving that function to local zoning appeals boards and to lower state and federal courts. Innumerable state appellate decisions on zoning policy and administration have since expanded upon the spare wording of *Euclid* and fleshed out the body of law commonly known as "Euclidean Zoning." The adoption of zoning by cities and towns nationally was encouraged by the 1924 Standard Zoning Enabling Act and its 1927 sequel, The Standard Planning Enabling Act developed by the U.S. Department of Commerce.[51] By 1929, over 650 local governments had adopted zoning and by 1939 virtually all cities had joined the zoning bandwagon (excepting the famous longtime holdout, Houston, Texas).

Regionalism: A Broader Template

With the blessing of the *Euclid* decision and the Department of Commerce model acts, America's cities and towns were turned loose to operate as independent fiefdoms for purposes of planning and zoning—and the property taxation that is closely interrelated with both. Despite efforts since the 1960s to "fix" zoning to serve various social, economic, and environmental objectives (discussed in chapter 4), most judicial pronouncements have addressed the *procedure* but not the *substance* of local zoning decisions. In particular, the administration of zoning by units of local government gives short shrift, if any consideration at all, to the needs of adjacent communities let alone the larger surrounding region.

The Supreme Court in *Euclid* acknowledged a tension between local and regional interests with the following little-noticed disclaimer: "It is not meant by this, however, to exclude the possibility of cases where the general public interest would so far outweigh the interest of the municipality that the municipality would not be allowed to stand in the way."[52] Nevertheless, the "interest of the municipality" has generally trumped a broader "general public interest." Metropolitan America became a vast mosaic of discrete zoning authorities operating under a set of rules established in the 1920s which have bedeviled attempts to promote a regional perspective ever since.

But in contrast to the legalistic, *micro*-scale perspective of early zoning advocates, another set of progressive urbanists viewed cities from the opposite end of the telescope, namely as *macro*-regions consisting of central cities, suburbs, and adjoining countryside. Rather than focusing on individual communities and parcels of land, the regionalists were wide-scale thinkers and, in some cases, effective doers. Little concerned about the separation of homes and businesses in neighborhoods (other than their own perhaps), the regionalists thought grandly and acted broadly. Their shared ideal was a future human habitat that blended commerce, industry, homes, open spaces, agriculture, and scenery—an orderly and efficient landscape blissfully free of "congestion," blight, and waste.

But regionalists differed vigorously as to how that ideal region would be achieved and what form it would take. Two leading factions centered in the New York area during the 1920s were the talented group of planners, architects, lawyers, engineers, and economists associated with the *Regional Plan of New York and Its Environs* ("The Pragmatists"), and those associated with the more utopian ideas of Ebenezer Howard, Patrick Geddes, and Lewis Mumford under the banner of the Regional Planning Association of America ("The Idealists"). Meanwhile, the interwar years spawned another strain of regionalism dominated by two architects of immense chutzpah—Frank Lloyd Wright and Le Corbusier ("The Prophets")—whose conflicting visions of future cities would each prove to be remarkably prescient.

The Pragmatists

The 1929 *Regional Plan of New York and Its Environs* was a mainstream, business-friendly initiative akin in pedigree, if not in its rhetoric, to the 1909 *Plan of Chicago*. Its two co-authors were Charles Dyer Norton and Frederic A. Delano—former members of the Commercial Club that developed the Chicago *Plan*—who returned to New York burning with zeal to apply their Chicago experience to the nation's biggest metropolis.

The New York regional study was breathtaking in geographic size and complexity— an area of 5,500 square miles extending forty miles beyond the city, containing parts of three states, over three hundred municipalities, and nearly ten million people. Norton's vision for this vast area was vintage progressivism: "To bring order out of disorder; to make convenience and thrift take the place of congestion and waste; and to realize the potentialities of commerce and of industry, as well as of beauty, and comfort, and pleasure."[53] (Note the downgrade of "beauty" to the "as well as" category.) According to the planning historian Harvey A. Kantor: "Norton had devised one of the most comprehensive ventures in the history of American city planning. The area he sought to encompass . . . was the largest yet attempted for this purpose and contained more people than any area of similar size in the United States. In addition, the personnel he proposed were the finest and most experienced in the nation. Norton was truly a giant among promoters of great planning for the New York area."[54]

Beginning about 1920, interest in the project among New York's patrician class gained momentum. Funding was committed by the Russell Sage Foundation (the sponsor of Lawrence Veiller's Tenement House Study two decades earlier). Key participants included Frederick Law Olmsted Jr., the attorney Edward Bassett, and Dwight W. Morrow, a J. P. Morgan partner, among other civic leaders. Several hundred attending the project's launch on May 10, 1922, were addressed by Secretary of Commerce Herbert Hoover and Elihu Root, a national *éminence grise* (and recipient of the Nobel Peace Prize in 1912). Norton proposed to inventory current conditions and recommend practical investments to promote the regional economy, notably rail and port facilities, highways, bridges and tunnels, water supply and sanitation, and public parks.

After Norton's untimely death in 1923, Frederick A. Delano (uncle of Franklin D. Roosevelt) became chair of the project with the Scottish town planner Thomas Adams as General Director of Plans and Surveys. The study was conducted by a blue-ribbon team including John Nolen and the urban planner Harland Bartholomew; the architects Cass Gilbert, Edward Bennett, and Charles A. Platt; and the landscape architect Frederick Law Olmsted Jr. Although Lillian Wald, founder of New York's Henry Street Settlement House, addressed the 1922 conference, she was not listed as an official participant in the study. Aside from secretaries, the New York Regional Plan team was emphatically male, mainstream, and moneyed.

After seven years of effort, the *Regional Plan of New York and Its Environs* appeared in ten volumes released between 1929 and 1931, with a total of 470 proposals. According to William H. Wilson: "The noblest of the era's comprehensive plans was the great *Regional Plan of New York and Its Environs*. 'Monumental' is a tired word but the only adjective adequate to this interdisciplinary effort by the outstanding urbanologists of the day. . . . The plan was a success in realistic terms, for many of its proposed highways, rail routes, parkways, and air terminals were built."[55] The *Regional Plan,* of course, did not begin with a blank slate. Projects in progress such as the George Washington and Triborough bridges, and various parks, highways, rail, and port facilities were assimilated into it. It also served as a blueprint for many depression-era projects directed by New York's rising technocrat-in-chief, Robert Moses.

The *Regional Plan* also yielded important nonstructural legacies. It helped to stimulate the spread of municipal and county planning authorities, and served as archetype for regional planning elsewhere around the country. It also demonstrated the potential for collaboration between the public and private sectors in planning and project implementation. Its most lasting institutional legacy was the formation in 1931 of the Regional Plan Association (RPA), which continues today under the leadership of Robert Yaro to provide a strong and objective voice for cooperative regionalism in the New York Region and elsewhere.[56]

In 1933, President Franklin Delano Roosevelt appointed his uncle Fred (Delano) to chair the new National Planning Board (NPB), with Harvard landscape architect Charles W. Eliot 2nd as executive director. Delano and Eliot spread the gospel of regional planning through many reports and conferences of the NPB and its successor agencies.[57]

The Idealists

The 1929 *Regional Plan for New York and Its Environs* did not meet with universal approval, however. Its business-friendly, economic trajectory offended those who were less satisfied with the status quo, especially as the depression took hold. It was criticized for its focus on transportation projects, especially for highway traffic, with only lip service paid to social issues such as affordable housing, sanitation, health care, workplace safety, and political corruption. Like the 1909 *Plan of Chicago,* it did not substantially advance the concerns of the settlement house movement and housing reformers like Catherine Bauer and Edith Wood.[58]

The New York *Regional Plan* was attacked by the acerbic Lewis Mumford in two 1932 *New Republic* articles for pandering to the New York business and professional elite: "No comprehensive planning for the improvement of living conditions can be done as long as property values and private enterprise are looked upon as sacred. . . . The Russell Sage planners . . . were so eager to fasten to . . . a solution acceptable to their committee full of illustrious names in financial and civic affairs, . . . that they deliberately restricted the area of their questions."[59]

With surprising bombast for someone still in his thirties, Mumford denounced the nation's foremost planning team for accepting "as 'automatic' and inevitable a process of metropolitan aggrandizement which has been in good part deliberate: the outcome of consciously formulated plans and purposes."[60] Furthermore, he exclaimed, "To assume that growth within an arbitrary metropolitan area will continue automatically in the future under the same conditions that prevailed in the past is to beg the whole question." And in case the point was still not clear: "[They] took the underlying causes of congestion as a 'given' and did not inquire into them or ask how they might be alleviated or diverted."[61]

Mumford thus posed the issue for American planners: Should their field simply "go with the flow" of urban development in further *centralizing* the growth of people and jobs in dominant urban complexes like New York? Or should planners seek to define and pursue a more normative future, namely, one of *decentralization*. Mumford in fact would be the self-appointed American apostle of "decentralization," the doctrine inspired by the charismatic English progressive Ebenezer Howard and his Scottish compatriot Patrick Geddes.

Howard was the last of the great nineteenth-century self-taught urban reformers (a tradition represented in America by Frederick Law Olmsted Sr., Jane Addams, and Jacob Riis, among others), and the first of a series of twentieth-century urban idealists.[62] Howard combined diverse strands of contemporary urban thinking and practice of his time into his concept of an ideal planned community, the "garden city."[63] Howard's famous "three magnet" diagram posed a Hegelian triad: *town* (offering jobs and culture at the expense of health, high prices, and crowding); *country* (a healthier environment at the price of boredom, poverty, and isolation); and as their synthesis, *town-country* (aka the garden city)—providing all of the advantages of town and country with no detriments whatsoever. Howard envisioned garden cities as free-standing, planned communities affording homes, jobs, shopping, recreation, and culture to their fortunate inhabitants. They would be interlaced with gardens and separated from one another and the larger metropolis by a "greenbelt" of farms and open spaces. Two prototypes were built outside London by Howard's disciples: Letchworth (ca. 1905) and Welwyn (ca. 1920).[64]

Geddes, the bearded Scottish botanist turned urban sage, enlarged the garden city perspective to the "level of the entire city region."[65] His inscrutable 1915 book, *Cities in Evolution*, envisioned harmonious coexistence of commerce, humans, and nature in the ideal metropolitan region of the future "with its better use of resources and population towards the bettering of man and his environment together . . . the creation, city by city, region by region, of its Eutopia [*sic*], each a place of effective health and well-being, even of glorious, and in its way unprecedented, beauty."[66] Anticipating the landscape architect [and fellow Scotsman] Ian McHarg by a half-century, Geddes sought to apply his botanical training to the interpretation of urban regions as akin to ecological organisms: "The case for the conservation of Nature, and for the increase of our accesses to her, must be stated more seriously and strongly than is customary. . . . On what grounds? In terms of the maintenance and development of life; of the life of youth, of the health of all. Such synoptic vision of Nature, . . . is more than engineering: it is a master-art; vaster than that of street planning, it is landscape-making; and thus it meets and combines with city design."[67]

The young Mumford eagerly embraced Geddes's eco-idealism and became his foremost American disciple and publicist. Their first meeting in 1923 led to the founding of the Regional Planning Association of America (RPAA), a small band of urbanists assembled by Mumford to publicize and realize the decentralist ideals of Howard and Geddes in America.[68] The urbanist Peter Hall has written that Geddes's "philosophy passed to a small, but brilliant and dedicated, group of planners in New York City, whence—through Mumford's immensely powerful writings—it fused with Howard's closely related ideas, and spread out across America and the world: exercising enormous influence, in particular, on Franklin Delano Roosevelt's New Deal, in the 1930s, and on the planning of the capitals of Europe, in the 1940s and 1950s."[69]

Closer to home, members of the RPAA designed several model housing developments inspired by garden city principles, beginning with Sunnyside Gardens in Queens, New York.[70] Radburn, New Jersey—dubbed by its promoters as "The Town for the Motor Age"—was the RPAA's sole attempt to build a garden city from the ground up. Eugenie L. Birch, a planning professor at the University of Pennsylvania, noted the benefits of their collaborative efforts: "The RPAA members worked as a team on the Radburn project and in the process they developed new planning methods. . . . In addition, the social scientists among them contributed significantly to the physical design."[71] The designers debated whether to admit African Americans and other nonwhites, and how to control future land use to protect Radburn's "garden" character.[72] Alas, the construction of Radburn by the CHC was suspended after the 1929 crash, leaving only three thousand dwelling units and a commercial center completed from the original RPAA plan.[73]

The "idealists" enjoyed one last fling: the quixotic New Deal Greenbelt Town program, directed by Rexford Tugwell. Tugwell envisioned Greenbelt Towns not as havens for factory workers out of industrial cities à la Howard and Mumford, but instead to provide opportunities for resettlement of the rural poor within reach of urban jobs.[74] Tugwell hoped that over the decades three thousand new towns nationwide would be built under the program; in reality, three small pilot towns—Greenbelt, Maryland; Greenhills, Ohio; and Greendale, Wisconsin—were started before the agency and its appropriation were judicially terminated after eighteen months. Peter Hall wryly sums up the impact of the program:

> In purely quantitative terms, . . . the greenbelt towns were almost a non-event: providing an attractive environment for only 2267 families can hardly be called that significant. . . . They are less important in fact for what they did, than for what they symbolized: complete federal control over development, bypassing local government altogether; . . . They provide, therefore, something of an exception in the first forty years of the garden-city movement. . . . There is a slight irony in that it all happened in the United States, which is almost the last country anyone would expect [federal community development] to happen. And there, it is hardly surprising that it failed.[75]

The Prophets

Standing apart from both the "pragmatists" and the "idealists" were two of the twentieth century's most influential and outspoken architects, Frank Lloyd Wright and Le Corbusier. Their respective visions of future cities differed radically from each other as well as from the other regional approaches discussed above. But their respective utopias proved remarkably prescient as templates for postwar American cities and suburbs.

Wright, a native of rural Wisconsin, famously hated cities despite his professional roots in Chicago and formative association with the eminent Louis Sullivan, "the genius of the Chicago School," as Hall describes him, "the man who gave definitive form to its greatest innovation, the skyscraper."[76] Sullivan and Wright mutually detested the Eurocentric neoclassicism of the Columbian Exposition and the Beaux-Arts craze it fostered. Wright launched his own practice in the 1890s designing suburban homes for wealthy clients, many in Oak Park where he built his own home and studio. His distinctive Prairie School homes, such as the Robie House at the University of Chicago, were low, horizontal structures that metaphorically hugged the ground with interior open floor plans, cantilevered rooflines, recessed entrances, and exterior walls of exposed brick, wood, and stone. Tiffany-style windows of beveled and stained-glass-bathed wood-paneled interiors with capricious shafts of sunlight and brilliant colors. Artwork, carpets, lamps, furniture, sometimes even table utensils, were meticulously prescribed by Wright in expressing his leitmotif—a blend of the Arts and Crafts decorative style and hints of Asian influence.

Wright's vision of Broadacre City as a "Usonian" (or utopian) future landscape reflected his conviction that existing cities should be dismantled and dispersed to liberate urban masses from the squalor of industrial cities. The electrical grid, automobiles, radio, telephone, and air travel would facilitate this process. Central to his thesis was the redistribution of land: like many progressives, Wright embraced the theories of Henry George that "the conflict between capital and labor was not fundamental to industrial society, and that this conflict, along with depressions and even poverty itself, could be eliminated if the root cause—inequities in land ownership—were eliminated."[77]

Broadacre City sprang in toto from Wright's imagination and hand in 1932 without reference to the work of Mumford's RPAA group. Where the garden city proponents sought communality and cooperation, Wright prescribed private ownership and independence. Instead of shared gardens and public greenbelts, he proposed allocating at least an acre of land to each family unit on which to live and plant gardens in blissful freedom with a short commute to nearby factory jobs and markets. Broadacre City failed the feasibility test in terms of his vision of land redistribution and the evolution of a family-centered, neo-agrarian, postindustrial paradise. One may scoff at his 1937 prediction that "Broadacre City is everywhere or nowhere. It is the country itself come alive as a truly great city."[78] But Broadacre City accurately predicted the coming dominance of the motor vehicle and the limited-access highway system in reshaping the face of the nation:[79] "Every Broadacre citizen has his own car. Multiple-lane highways make travel safe and enjoyable. There are no grade crossings nor left turns on grade. The road system and construction is such that no signals nor any lamp-posts need be seen. No ditches are alongside the roads. No curbs either."[80]

In stark contrast to Wright's decentralized Usonian future, the French architect Le Corbusier proposed the exact opposite: high-density, new cities planned and administered by central authorities. Robert Fishman, a professor of architecture and urban planning at the University of Michigan, observes that such diametrically opposed solutions evolved from identical premises: "Wright and Le Corbusier seem predestined for comparison. Their ideal cities confront each other as two opposing variations on the same utopian theme. Both believed that industrialization had produced the conditions for a new era of justice, harmony, and beauty; that this era would commence with the replacement of existing cities by new forms of community suited to the new age; and that this physical restructuring of society would be the fundamental revolutionary act separating the past from the future."[81]

Wright and "Corbu" also shared the conviction that "physical restructuring of human society" was essentially an architectural problem to be solved through application of appropriate technical and design principles. Issues of economy, demography, and governance were secondary to the realization of the ideal physical setting. Their respective architectural vocabularies determined the character not only of their buildings but also of their proposed utopias—cities as buildings writ large. Wright's predilection for "earthy" materials—wood, brick, and stone— underlay the "back to the land" horizontality of Broadacre City. Conversely, Corbu's ideal city was a vertical phalanx of concrete, steel, and glass towers set amid open spaces. His signature *Plan Voisin* of 1925 imagined the demolition and rebuilding of central Paris, with the allocated towers to offices and the populace assigned to satellite clusters of apartment blocks and "garden cities." In his 1935 design for *Ville Radieuse* (The Radiant City), the central towers are residential, with business relegated to less auspicious buildings. Apartments in the central towers "are not assigned on the basis of a worker's position in the industrial hierarchy but according to the size of his family and their needs."[82] Four-fifths of the total project site was to be devoted to parks, gardens, playing fields, cafes, and outdoor markets. The dominance of the city street was to be reduced by limiting vehicular traffic to superhighways leading to and from the city, and to internal roadways physically separated from pedestrian walkways. Like Victor Gruen's shopping malls of the 1960s, the central management allocated all space; nothing was left to the whim of the marketplace.

Postwar America borrowed heavily from the utopian playbooks of both Wright and Le Corbusier in the layout of suburbs and the "renewal" of central cities, as discussed in the next two chapters. But this borrowing only utilized the bare forms of Broadacre City (the horizontal plan of banal single-family homes, highways, and malls) and *Ville Radieuse* (the vertical plan of vacuous high-rise apartment blocks)—caricatures of utopias gone sour. Such, perhaps, is the fate of architects with the hubris to prescribe how human society should live and function.

Pitching the Future: Worlds Fairs of the 1930s

World's fairs were a remarkable staple of the depression, during which six American cities hosted such events: Chicago in 1933–34; San Diego, 1935; Dallas, 1936; Cleveland, 1936–37; San Francisco, 1939–40; and New York City, 1939–40.[83] Although their themes, size, and economic success varied widely, the six fairs collectively projected a vision of a new world, new cities, and a new prosperity and quality of life based on technology. By the far the largest and most influential of the fairs were those in Chicago and New York.

Chicago's Century of Progress International Exposition, commemorating the city's centennial, was conceived in the heady atmosphere of the late 1920s, before the stock market crash, but planning and construction went ahead nonetheless. In spite of the depression, the 1933–34 event set new records, with 48.7 million visitors over two seasons,[84] and generated an estimated $770 million in economic benefits to hotels and other city businesses.[85] (The lakefront fair site was later used for a commuter airport and finally, at the behest of Mayor Richard M. Daley in the mid-2000s, was converted to park use, as visualized in the 1909 *Plan of Chicago*.)

The Century of Progress fair was an orgy of art deco and modernist architecture, spectacular lighting, and post-Repeal revelry. The equivalent to the Ferris wheel of 1893 was the Sky Ride, with two 625-foot towers topped by observation decks from which "rocket cars," suspended 225 feet above the ground, traversed the width of the fairgrounds.[86] Frank Lloyd Wright's work was not represented, but the fair's architecture showcased the modernism then emanating from Germany as well as from Chicago itself. With City Beautiful–era ornamentation largely banished, excitement and color were provided through dazzling lighting effects: "In addition to lighting up the fair buildings, the designers created other impressive effects with spotlights and the fountains. A bank of twenty-four arc lights . . . radiated a virtual aurora borealis above the Fair that could be seen for miles."[87]

Six years later, New York City held the most ambitious (albeit not the largest) world's fair ever whose sanguine title "World of Tomorrow," ironically belied the breakout of World War II. At the local level, political showboating and one-upmanship plagued the planning; James Mauro, in *Twilight at the World of Tomorrow*,[88] recounts the near disasters of mounting such an enterprise. Like Chicago's

Century of Progress, New York's World of Tomorrow would occupy a vast tract of reclaimed land, a former garbage dump amid the Flushing Meadows wetlands near LaGuardia Airport in Queens. Robert Moses, czar of the city and state park systems at the time, directed the leveling of the site and its preparation for the fair in the expectation that it would subsequently be a huge new park to serve the rapidly growing population in Queens, Brooklyn, and the Bronx.[89]

As the historian Robert Rydell has noted, all six of America's depression-era world's fairs envisioned a better future. "These fairs, individually and collectively, helped Americans find their way out of the Great Depression. They did so not by accident, but by design. Attended by some one hundred million visitors, these fairs left in their wake a vision of the United States emerging from the Great Depression as a consumer-centered, corporation-driven nation-state powered forward by science and technology and governed by a federal government made newly attentive by the circumstances of the 1930s to the welfare of its citizens."[90]

Chicago's Century of Progress featured pavilions of radical design erected by the giants of the burgeoning automobile industry: General Motors, Chrysler, Ford, Firestone Tire, Owens-Illinois Glass, Standard Oil, Havoline Oil. Chrysler provided a quarter-mile test track where "Hell Drivers" demonstrated how not to drive in the real world. Not to be outdone, the General Motors pavilion included an entire Chevrolet assembly line that rolled out real cars for sale. Henry Ford sat out the first season, but in 1934 he "stole the show by building the largest and most expensive corporate pavilion at the fair," designed by the eminent modern architect Albert Kahn, and equipped with a driving track where visitors could try out the latest Ford models.[91] The fair also featured model homes designed for the middle class by leading architects, each low cost, single family, and suburban in style. All in all, Century of Progress convinced millions that America's post-depression future would be spent riding in classy roadsters to and from a bungalow in the suburbs. They were correct, at least for the white middle class.

The New York fair added more automotive hoopla, a vision of the cities and suburbs where "modern" Americans would live and the highways on which they would travel between them. Two immensely popular pavilions, Democracity and the General Motors (GM) Futurama, presented exquisitely detailed model cityscapes (fig. 2.5) Democracity, located within the spherical half of the fair's iconic Trylon and Perisphere structures, was created by the celebrated industrial designer Henry Dreyfuss (who also designed the streamlined "20th Century Limited" train and many household products). His fantasy metropolis set in the year 2039 echoed some themes of Wright's Broadacre City but with more technical detail and special effects. James Mauro, in his history of the fair, describes Democracity as "a grand diorama that depicted a utopian landscape of the future—a dreamy vision of a world in which people lived in countrified splendor while having easy access to urban industry via convenient planning and a broad highway system. Five satellite

Fig. 2.5. General Motors Futurama pavilion at the 1939 World's Fair. Harry Ransom Center, The University of Texas at Austin. Image courtesy of the Edith Lutyens and Norman Bel Geddes Foundation.

towns (including 'Pleasantville' which eerily predicted the suburb of the future) surrounded 'Centerton' the industrial core. For the overcrowded city dweller as well as the isolated farmer, Democracity represented the ideal promise of a better world they so eagerly craved."[92]

The General Motors Futurama, designed by Norman Bel Geddes, also dazzled its millions of visitors with a brave new world set in 1960. Futurama viewers sat in moving chairs with individual speakers high above an ever-shifting images of skyscraper cities and idyllic suburbs crisscrossed by limited-access highways traversed by thousands of miniature Buicks and Chevrolets (presumably occupied by smiling white families): "Both exhibits," the historian Robert Bennett has remarked, "displayed modes of decentralized skyscraper cities that were thinly veiled variations on the abstract, geometrical modernism of Le Corbusier's *Ville Radieuse* and *Plan Voisin*" and extended Corbu's "ideal of the modernist home as a 'machine for living in' across entire megalopolitan regions, turning the landscape itself into a vast, finely tuned, and intricate machine."[93]

The *New Yorker* columnist E. B. White later recalled the GM pavilion with his customary wry skepticism:

> The countryside unfolds before you in five-million-dollar micro-loveliness. . . . The [narrator's] voice is a voice of utmost respect, of complete religious faith in the eternal benefaction of faster travel. The highways unroll in ribbons of perfection through the fertile and rejuvenated America of 1960—a vision of the day to come, the unobstructed left turn, the vanished grade crossing, the town which beckons but does not impede, the millennium of passionless motion. . . . I liked 1960 in purple light, going a hundred miles an hour around impossible turns ever onward toward the certified cities of the flawless future. . . . [However,] the apple tree of Tomorrow, abloom under its inviolate hood, makes you stop and wonder. How will the little boy climb it? Where will the little bird build its nest?[94]

More bluntly, the celebrated columnist and political commentator Walter Lippmann questioned the underlying imperative of the futurist exhibits, that government should remake the nation in a form more profitable to the car industry: "GM has spent a small fortune to convince the American public that if it wishes to enjoy the full benefit of private enterprise in motor manufacturing, it will have to rebuild its cities and highways by public enterprise."[95]

In contrast to the metropolitan images of Futurama and Democracity, Mumford and his RPAA colleagues weighed in at the 1939 fair via the film *The City*, produced by the American Institute of Planners with a musical score by Aaron Copland. The film opens with a glimpse of bucolic Shirley, Massachusetts (home of RPAA member Benton MacKaye), where arcadian harmony reigns. The scene shifts to smoky, noisy, overcrowded Manhattan, the antithesis of Shirley. Inevitably, the proposed synthesis is the garden city represented by Tugwell's showcase, Greenbelt, Maryland. In Mumford's words, emblazoned on the screen: "Year by year, our cities grow more complex and less fit for living. The age of rebuilding is here. We must remold our old cities and build new communities better suited to our needs."

While the intelligentsia thus joined battle over conflicting visions of American utopias, these debates barely touched on the question of race: everyone concerned seemed to agree implicitly that the future belonged to the white middle class. Racism was pervasive in all aspects of the 1930s-era world's fairs. In Chicago, blacks were hired only as grounds laborers or maids in the model home exhibit. In Dallas, a "Hall of Negro Life" was included under pressure from the federal government (Eleanor Roosevelt perhaps?), but no African Americans were engaged to design it.[96] And on opening day of the 1939 New York fair, blacks demonstrated against racial discrimination in hiring. A "Negro Week" in the second season of the New York fair divided black leadership on the question of the wisdom of participating.[97]

Visitation to the fairs by African Americans was discouraged in effect, if not officially, through the prevalence of Jim Crow segregation. The World of Tomorrow would be modernist in architecture, transportation, household appliances, furniture, clothing, and hairstyles but profoundly reactionary in its racism.

The Patrician Decades in Retrospect

The first three decades of the twentieth century witnessed intense and exuberant city growth in population size, wealth, and skyline—offset however, by chronic fault lines between rich and poor, men and women, white and nonwhite, and native and foreign-born. The share of Americans living in the nation's ten largest cities peaked at 15.1 percent in 1930, and declined over the rest of the century to 8.5 percent in 2000.[98] With the shift from horse-drawn streetcars to electric commuter rail and automobiles in the 1910s, independent suburbs gradually enveloped the older industrial cities, blocking their outward expansion. But by 1930, suburbanites still numbered fewer than half the population of central cities. (By the century's end that ratio would be reversed.) The skyward growth of the nation's business districts continued after the 1929 crash—the 102-story Empire State Building was completed in 1931 but had few tenants (and was quickly dubbed the "Empty State Building").

With the advent of the New Deal in 1933, cities for the first time were showered with federal funds to create jobs and build bridges, tunnels, parks, housing, and other infrastructure. The six depression-era world's fairs were strongly supported by Congress and the White House to boost morale, provide jobs, and showcase new federal and corporate urban priorities.

Cities were hobbled by limits on the legal powers granted to them by states, and by corrupt "political machines" like New York's Tammany Hall. But many cities, large and small, were endowed with energetic citizens, public-spirited philanthropists, thriving local support for urban improvements, and a growing cadre of civil engineers who built or expanded critical urban lifelines—water and sewer systems, railroads, bridges, tunnels, highways, schools, subways, communications, electrical and gas facilities—many of them still in use today. These decades also nurtured the minds, spirit, and health of the growing middle class through the building of universities, libraries, museums, concert halls, parks, and hospitals, often funded by wealthy benefactors. Also playing a vital role were religious institutions whose churches, synagogues, temples, and the occasional mosque uplifted the visual and spiritual life of cities and their neighborhoods.

As recounted above, the settlement house movement and the City Beautiful movement, both emerging from Chicago at the turn of the century, offered contrasting approaches to urban self-betterment. Elements of each of those movements in turn contributed to the rise of the city planning movement in the 1910s and 1920s. The various regional scenarios of the 1930s sought to enlarge the planners' quest for order and efficiency to the metropolitan scale. Ironically, the foremost achievement of the early city planning era—local land use zoning—would prove to be the enduring nemesis of regionalism and a mockery of the intentions of its early supporters, as I discuss in chapters 4 and 6.

Nevertheless, the Patrician Decades were a hopeful time of trial and error in the art of making cities more functional, healthy, and safe (if not equitable). The creative tension between the settlement housers and the City Beautiful partisans presaged a century of struggle between top-down technocracy and grassroots social and environmental advocacy. But those well-meaning partisans early in the century at least held a common set of five progressive articles of faith:

1. *Spatial determinism*—the evils of "congestion, disorder, and inefficiency" arise from and reflect deficiencies in the physical layout of cities, their circulatory systems, their land use practices, and lack of open space.
2. *Public intervention* in the private economy is essential to protect the public health, safety, and welfare from the abuses and hazards of unbridled laissez-faire.
3. *Key urban infrastructure* (e.g., highways, bridges, schools, water and sewer systems, parks) must be provided by responsible public authorities and funded through user fees or taxes, according to the service provided.
4. *Research*—careful documentation and analysis of urban problems must be the basis on which public policies are based.
5. *Public support* for progressive proposals may be galvanized through such media as journalism, conferences, films, political lobbying, and, when necessary, mass demonstrations.

PART II

The Technocrat Decades, 1945–1990

The year 1945—with both V-E Day (May 8) and V-J Day (September 2)—brought to an end the fifteen-year nightmare of the Great Depression and World War II. The spirit of the moment was captured in Alfred Eisenstaedt's iconic photograph of a sailor kissing a girl in the middle of Times Square. Who realized or cared that Europe had been carved up yet again, this time dividing "West" and "East" by a boundary—soon termed the "Iron Curtain" by Winston Churchill—that would be the political fault line of the cold war? And why worry about the implications of dropping atomic bombs on two Japanese cities, portending the possible demise of civilization? It was time to celebrate and then to build houses and start families.

Those fifteen years of combating depression and despotism had irreversibly expanded the United States government and changed its relationship to American society. Before the 1929 stock market crash, the nation's domestic well-being was largely the concern of states, local governments, and the private sector. After the New Deal and the war, the federal government, for better or worse, had become the dynamo of domestic policy, power, and funding, with other sectors reduced to accessory roles. Nowhere would this be more dramatically apparent than in the case of housing and transportation policies that epitomized the "Technocrat Decades" with their combined imprint on the geographies of American cities and suburbs. In the immediate postwar period, business-led coalitions committed to a capitalist economy urged the federal government to finance (on a regular basis, not just in emergencies) programs that supported growth and spurred further investment.[1]

Housing was the first order of business. With fourteen million young adults demobilizing from the armed forces in 1945 and eager to "settle down," there was a desperately short supply of suitable housing. According to one estimate, the nation needed at least five million new homes and apartments to satisfy the immediate demand.[2] Reflecting the new dominance of the federal government, demands by politicians and the media for new homes were directed not to city halls or statehouses but to Congress. Even such a conservative business magazine as *Fortune* "published dozens of articles in 1946 and 1947 on the housing shortage," the environmental historian Adam Rome has noted: "In a rare editorial—'Let's Have Ourselves a Housing Industry'—the editors supported a handful of government initiatives to encourage builders to operate on a larger scale . . . as the best defense against socialism."[3] That view was not unanimous, however. In 1947 and 1948, a brash young senator from Wisconsin, Joseph McCarthy, tried to block federal participation in housing (especially multifamily housing) programs as "communistic."[4]

Congress responded to the housing shortage with a two-track set of postwar national housing policies and programs, effectively legislating an American version of apartheid. The first track, which I discuss in detail in chapter 3, began with a New Deal program for slum clearance and public housing for the poor, which was soon transcended by all-out redevelopment of vast swathes of urban America through the dual sledgehammers of urban renewal and interstate highway construction. Low-income African Americans dislocated from areas declared "blighted" or proposed as highway corridors either ended up in desolate public housing or added to overcrowding in other tenement districts since they were not allowed to join the white exodus to the suburbs.

The second track, the topic of chapter 4, launched a crash program of federal support to the housing industry to build single-family homes for the white middle class outside older cities. Two pre-war agencies, the Federal Housing Administration (FHA) and the Veterans Administration (VA) were charged with jump-starting housing construction by stimulating both supply and demand—the former through FHA loan guarantees on behalf of eligible builders, and the latter through zero-down-payment, low-interest thirty-year loans to veterans and other qualified borrowers. These loan guarantees, as well as new federal highways, tax incentives, and other federal subsidies fueled a construction boom of some fifteen million new housing units during the 1950s; in every year from 1947 to 1964, housing starts would exceed 1.2 million.[5] Most of these millions of new units were single-family homes built on agricultural or wooded land outside of the older central cities. Most were marketed exclusively to whites.

This dual set of housing policies—both very agreeable to the building industry and to local governments—reflected a set of unstated but powerful axioms of American society and politics in the 1940s, vestiges of which endure today:

- *Racism.* Pre-war racist attitudes and policies continued to flourish after the war despite the years of shared military sacrifice and home-front burdens. In 1944 the Swedish sociologist Gunnar Myrdal wrote: "Except for a small minority enjoying upper or middle class status, the masses of American Negroes . . . are destitute. They own little property, their incomes are not only low but irregular. They thus live from day to day and have scant security for the future."[6] The housing boom that began in the mid-1940s was accordingly dichotomous: whites and blacks would fare very differently under public policies and private practices that ensured separate, and very unequal, access to new housing opportunities.
- *Privatism.* While the New Deal greatly expanded public intervention in the private economy—with Social Security, banking regulation, collective bargaining, farm price supports, and home mortgage insurance—the nation emerged from the depression and World War II supremely confident that capitalism could solve any problem, including the housing shortage. Federal subsidies and incentives were welcome, but direct federal housing construction in competition

with private enterprise was anathema to the business sector and its political representatives in Congress.

- *Localism.* Described by Jon C. Teaford as "a sacred element of the American civil religion,"[7] local government regained its hallowed status in the 1940s after the brief flirtation with "regionalism" during the previous decade. The mosaic of central cities enveloped by suburban jurisdictions was firmly established by then. While cooperating selectively on regional needs such as water and sewer services, suburbs were implacably hostile to the central city in matters of housing, education, and taxes.

- *Federalism.* Notwithstanding the strength of privatism and localism, the expansion of the federal government during the depression and the war was irreversible. Even though it had been unthinkable as recently as the 1920s, all three branches of the federal government were inextricably involved in the nation's economy and social order by the late 1940s. It was inevitable therefore that Congress be called on to contribute to efforts to expand the general housing supply while eliminating slums in cities. But responsive to the conservatives, this contribution would support and enable localities and the building industry to remake urban America as profitably and inequitably as they wished.

- *Elitism.* A legacy from the City Beautiful and planning movements earlier in the century, it was conventional wisdom in the 1940s and 1950s that communities and neighborhoods should be planned or reconstructed by higher authorities (governmental and corporate). The wishes of residents of neighborhoods proposed for redevelopment were not important, since they would almost certainly move somewhere else. Likewise, the interests of future subdivision residents were not reflected, since they had not yet actually arrived. The fate of older city neighborhoods and new bedroom communities alike rested in the hands of local officials, developers, realtors, and lenders who relied on professional design consultants. Those consultants, moving quickly from one client to another, tended to recommend "boilerplate" plans, with the result that postwar construction for both the white and the black housing markets was dull and stultifying—and much of it is still around.

- *Modernism.* The dominant architectural paradigm of the postwar period represented order, efficiency, uniformity, and power (governmental or corporate) to the detriment of flexibility, diversity, and individuality, according to critics such as Lewis Mumford, Catherine Bauer Wurster, and Jane Jacobs.

Chapter 3

The Central City Renewal Engine

> *But look what we have built with the first several billions [of urban renewal funds]: Low-income projects that become worse centers of delinquency, vandalism and general social hopelessness than the slums they were supposed to replace. Middle-income housing projects which are truly marvels of dullness and regimentation, sealed against any buoyancy or vitality of city life. . . . Expressways that eviscerate great cities. This is not the rebuilding of cities. This is the sacking of cities.*
>
> —JANE JACOBS, *The Death and Life of Great American Cities,* 1961

"Urban renewal" is a toxic phrase. Broadly speaking, urban renewal encompassed an array of federal, state, and local programs that collectively sought to eliminate slum districts; construct new housing and commercial development under private and public auspices; shore up urban tax bases; and stimulate private investment in the vicinity of project areas. More narrowly, the term refers to the program of federal assistance to local renewal agencies established in Title I of the 1949 Housing Act. Once the mantra of powerful city rebuilders such as Robert Moses in New York, Edward Logue in New Haven and Boston, and Edmund N. Bacon in Philadelphia, the term and concept soon became an epithet to detractors like Jane Jacobs, William H. Whyte, Herbert J. Gans, Charles Abrams, and Joseph Fried (who wrote memorably that two decades after the program's inception in 1949, the term "was spat bitterly from the mouths of slum dwellers as often as it was rolling pridefully from the tongues of mayors and chamber of commerce officials").[1]

By the late 1960s, hundreds of projects were in progress, but the concept as originally conceived in the 1940s was thoroughly discredited on many counts. It was damned by its critics for its excessive clearance of viable buildings and neighborhoods, its tilt toward upscale redevelopment in desirable locations, its backwash of vacant lots and shoddy public construction in "downscale" areas, and its abject failure to provide decent replacement housing and commercial space for the households and businesses evicted from project sites—a phenomenon worsened by the federal highway program.

The road to the hell that urban renewal became was paved, of course, with good intentions. Two of its well-meaning antecedents were the New Deal public housing program, the beau ideal of progressive housing advocates that became their bête noir in its postwar reformulation; and Stuyvesant Town, a phalanx of high-rise, middle-class apartment buildings on the East Side of Manhattan conceived during the war to house returning veterans, and which remains today a conflicted but vital element of New York's housing stock.

Public Housing: A "Dreary Deadlock"

The Rube Goldberg edifice of federal urban renewal rested uneasily on the precedent of federal assistance to local housing authorities to build and manage public housing for the very poor. Such efforts began modestly with the funding of about 22,000 low-rent units in fifty projects by the Public Works Administration between 1934 and 1937.[2] In the latter year, the Wagner-Steagall Housing Act established a new program to assist local authorities to build and operate housing for "families who are in the lowest income groups and who cannot afford to pay enough to cause private enterprise . . . to build an adequate supply of decent, safe, and sanitary dwellings for their use."[3]

The 1937 act was the fruit of years of advocacy by progressive housing reformers, most notably Catherine Bauer (later known by her married name, Wurster), Edith Elmer Wood, and members of the settlement house movement. Unlike Lawrence Veiller, who earlier in the century had advocated stronger building regulations to promote better low-cost housing, Bauer, Wood, and their fellow "housers" sought federal funding for construction and management of new low-rent housing by local public agencies. President Franklin D. Roosevelt endorsed their goals in his second inaugural address, in which he declared, "I see one-third of a nation ill-housed, ill-clad, ill-nourished."[4]

Between 1937 and 1941, the U.S. Housing Authority, established by the act and directed by Bauer, sponsored some 130,000 units in about three hundred projects across the country.[5] Most of these featured low-rise garden apartments amid green open spaces, reflecting the influence of the Regional Planning Association of America, in which Catherine Bauer was a charter member (as well as confidante of its primary founder, Lewis Mumford). Separate projects served both white and black tenants, including working-class households (fig. 3.1).

Building a "Social Chernobyl"

The idealism of New Deal–era public housing did not survive into the postwar period. Although "slum clearance" was not its explicit purpose, urban officials often approached public housing with only grudging attention to the need to re-house people displaced from clearance areas. Furthermore, under the forceful direction of Robert Moses, the New York City Housing Authority (NYCHA) led the national trend away from "garden-style" public housing toward high-rise barebone structures largely occupied by very poor African Americans. And except for housing projects for the elderly, most were constructed in districts already solidly nonwhite and poor. Moses incorporated decent landscaping, play areas, and

Fig. 3.1. Low-rise pre-war public housing units built under the 1937 National Housing Act, Holyoke, Massachusetts.

community facilities in early projects built under city and state auspices. But, to placate postwar conservatives such as McCarthy, federal authorities imposed tight limits on per unit costs and eligible household incomes to ensure that public housing would serve only the very poor and thus not compete with the private sector. In the case of New York City, high site costs forced Moses and the housing authority to eliminate all "frills" and build only barracks-like high-rise towers separated by dreary patches of grass—a dark parody of Corbusier's "towers in a park."[6]

By 1952, NYCHA housed 265,000 people in sixty-eight projects built under city, state, and federal auspices. But slum clearance far outpaced the supply of new public housing units. According to the urban historian Nicholas Dagen Bloom, "By 1957 . . . *all* NYCHA projects built by that time had been able to house only 18.1 percent of [clearance] site tenants." Displaced households were paid a modest compensation and evicted, leading to overcrowding of other nearby tenements. This cavalier attitude by Moses and the NYCHA "reflects the technocratic, insensitive blind spot of the authority that allowed it to destroy whole neighborhoods."[7] By the mid-1960s, with the critics now raising a hue and cry, NYCHA public housing reached its peak with a half million very poor, mostly black tenants in 154 projects. (Robert Moses had been ousted as public housing czar in 1959 but continued to direct most of the city's public works programs until the late 1960s. See "Robert Moses: New York's Nemesis, Savior, or Both?" below.)

By 1970, some 870,000 units of public housing had been constructed under the 1937 housing act, scattered among cities large and small in all fifty states.[8] Most of these units were built in large cities, however, where the program's failures have been most notorious. Many of the Moses-era projects in New York managed to survive into the present century, much of their housing stock renovated and functional today.[9] Other cities, however, have had to terminate their ill-fated experiments with high-rise Corbusian modernism and begin over again. The nation's most infamous public housing fiasco was the Pruitt-Igoe project in St. Louis. Completed in 1955 with thirty-three eleven-story buildings and 2,868 apartments, Pruitt-Igoe became a metaphor for warehousing black welfare families in filthy, noisy, crime-ridden enclaves isolated from jobs, schools, parks, and public transportation. Nearly abandoned by even its most needy tenants, Pruitt-Igoe was demolished in 1972, once the city and housing authority acknowledged that it comprised a physical and social monstrosity beyond redemption.

Similarly, the Chicago Housing Authority earned ignominy for its many high-rise and dysfunctional projects in predominantly black districts of the South and West Sides of the city. Among five projects along State Street ("that Great Street," as the song goes) south of the Loop was the reviled Robert Taylor Homes: twenty-eight sixteen-story apartment buildings abutting the new fourteen-lane Dan Ryan Expressway that, in one critic's words, "served as a practically impenetrable wall between State Street public housing and the all-white neighborhoods on the other side."[10] The housing experts Paul S. Grogan and Tony Proscio described the status of public housing in Chicago and other cities in the 1950s and 1960s as "a social Chernobyl, stifling the lives within and radiating blight on the neighborhoods beyond. Nearly everything about them seemed engineered for self-destruction: irrational tenant selection and eviction rules, hideous design, neglectful maintenance, and a deliberate concentration of poverty that Blaine Harden of the *Washington Post* described as 'stacking poor people in human filing cabinets.'"[11]

When I visited it as a graduate student in 1970, Chicago's Robert Taylor complex was a wasteland of disintegrating buildings, sodden, empty play spaces, "temporary" mobile classrooms, and disabled cars (fig. 3.2). The adjacent Dan Ryan Expressway exposed the tenants to noise, pollution, and visual ugliness while giving no access to the roadway or to the rapid transit line in its median strip. The residents of Robert Taylor were effectively cut off from the city that surrounded them. Arguably, living conditions there were more atrocious than in the former walk-up tenements that were cleared for both the project and highway. (In 1999, the Chicago Housing Authority at the behest of Mayor Richard M. Daley began to demolish all of its high-rise public housing projects and to replace them with

Fig. 3.2. "Social Chernobyl": an unusable playground at the Robert Taylor Homes public housing project, Chicago, ca. 1970.

mixed-income, low-rise housing scattered across the city. The two-mile stretch once occupied by Robert Taylor Homes is now Legends South.[12]

The dismal postwar public housing program was soon disowned by Catherine Bauer, a champion of the 1937 Wagner-Steagall Act. In an *Architectural Forum* essay titled "The Dreary Deadlock of Public Housing," Bauer lamented: "Public housing, after more than two decades, still drags along in a kind of limbo, continuously controversial, not dead but never more than half alive." The program was widely despised by private builders, lenders, and property owners and commanded no support from "civic-minded groups." In addition to the "rigid and heavy-handed" management by local housing authorities, Bauer cited two fatal blunders: the formulation of public housing as a separate program unrelated to other metropolitan housing policies and the infatuation of architects and planners with Corbusian "modern design" as the universal norm for housing the very poor. High-rise towers, she perceived, "whose refined technology gladdens the hearts of technocratic architectural sculptors leaves no room for personal deviation, for personal initiative and responsibility, for outdoor freedom and privacy." Moreover, "management domination is built in, a necessary corollary of architectural form."[13]

Federal public housing thus provided a hapless precedent for the much larger Title I urban renewal program. Despite the best intentions of the 1937 Housing Act

Robert Moses: New York's Nemesis, Savior, or Both?

Robert Moses (1888–1981) personified the "Technocrat Decades" as one of the most powerful and controversial *unelected* public officials in U.S. history. According to his biographer Robert A. Caro, he out-powered every New York mayor and state governor over four decades.[1] From the 1920s into the 1960s, Moses dominated public works construction in metropolitan New York. As "the best bill drafter in Albany,"[2] he established and controlled a wide-ranging empire of special authorities, commissions, and boards whose hub was the Triborough Bridge and Tunnel Authority, his headquarters and toll-based cash cow. By the early 1960s, Moses chaired twelve different public authorities.

His political reach extended to the White House and Congress—from FDR's New Deal through LBJ's Great Society—with profound influence over federal housing, urban renewal, and highway programs. According to Moses's *New York Times* obituary by Paul Goldberger, "the works he created in New York proved a model for the nation at large. His vision of a city of highways and towers—which in his later years came to be discredited by younger planners—influenced the planning of cities around the nation."[3] Caro wrote of him: "He was the greatest builder in the history of America, perhaps in the history of the world."[4]

A descriptive catalogue of "Built Work and Projects" overseen by Robert Moses in New York City alone fills 180 pages of a recent reappraisal of his legacy, edited by the urban historians Hilary Ballon and Kenneth T. Jackson.[5] The quantity, quality, and diversity of projects which he built or strongly influenced ranged from about seven hundred playgrounds and dozens of public pools and other neighborhood recreation facilities to the spectacular Bronx-Whitestone, Throgs Neck, and Verrazano-Narrows bridges, the Brooklyn-Battery Tunnel, Lincoln Center, the United Nations, two world's fairs (1939 and 1964), and untold housing, urban renewal, and highway projects. His pre-war state parks and parkways on Long Island, anchored by world-famous Jones Beach State Park, comprise a justly admired triumph of regionalism.

Far less appreciated, both at the time and in hindsight, were his postwar public housing and urban renewal projects and highways slashing through the outer boroughs. His proposed extension of Fifth Avenue through Washington Square Park and plans for the ten-lane Lower Manhattan Expressway sparked intense opposition spearheaded by Jane Jacobs, which led to their defeat.[6]

Deeply troubling to later critics, influenced perhaps by Robert Caro's Pulitzer Prize–winning book *The Power Broker: Robert Moses and the Fall of New York,* were the bare-knuckle tactics Moses used to influence or demolish political opponents. Similarly, "the heartbreaking callousness with which Moses evicted tens of thousands of poor people in his way"[7] became a lasting stain on his legacy. Moses has long been recognized as unapologetically racist, manifested in his eviction practices, his segregated housing projects, and his bias on behalf of highways for the (white) middle class over public transportation serving the broader urban populace.

In his reconsideration of the Moses legacy published in 2008, Kenneth Jackson concedes that the Master Builder was unquestionably racist, but argues that this was unfortunately consistent with the spirit of his time and must be balanced against what he accomplished:

Moses had contempt for the poor and rarely expressed admiration for African-Americans. But he did have a consistent and powerful commitment to the public realm: to housing, highways, parks, and great engineering projects that were open to everyone. While Moses was in power, the word "public" had not yet become pejorative, and the power broker was willing to override private interests in order to enlarge the scope of public action. In the twenty-first century, when almost anything "public" is regarded as second-rate, and when the city cannot afford to repair—let alone construct—grand edifices, that alone is a remarkable achievement.[8]

Indeed, as time passes, some of Robert Moses's visions have proven astoundingly prescient. For instance, Moses consented to apply his mighty city public works resources to reclaim an immense dump in Queens as the site of the 1939 World's Fair. Although he was (rightly) skeptical that the fair would succeed financially, he visualized that the site would one day become "a park to end all parks, one and a half times the size of Central Park, that would be known as 'the Versailles of America.' "[9] In due course, after its reuse for the 1964 fair, the site was transformed into Flushing Meadows Corona Park. In 2011, the *New York Times* reported that the park—now flanked by the new Mets Stadium, the National Tennis Center, Queens College, and the Queens Zoo, Botanic Garden, and Museum of Art—is today a vast outdoor pageant of sports, food vendors, music, and festivals, serving the Chinese, Korean, Mexican, Ecuadorian, and Colombian populations of nearby communities.[10]

1. Robert A. Caro, "The City Shaper," *New Yorker,* January 5, 1998, 38.
2. Robert A. Caro, *The Power Broker: Robert Moses and the Fall of New York* (New York: Knopf, 1974), chap. 10.
3. Paul Goldberger, "Robert Moses, Master Builder, Is Dead at 92," *New York Times,* July 30, 1981.
4. Caro, *Power Broker,* 43.
5. Hilary Ballon and Kenneth T. Jackson, eds., *Robert Moses and the Modern City: The Transformation of New York,* ed. (New York: W. W. Norton, 2007).
6. Jacobs's near-sainthood in the realm of urban planning has been reaffirmed by two recent books that revisit the Jacobs-Moses struggle: Roberta Gratz, *The Battle for Gotham: New York in the Shadow of Robert Moses and Jane Jacobs* (New York: Nation Books, 2010), and Anthony Flint, *Wrestling with Moses: How Jane Jacobs Took on New York's Master Builder and Transformed the American City* (New York: Random House, 2011).
7. Caro, *Power Broker,* 42.
8. Ballon and Jackson, *Robert Moses and the Modern City,* 70–71.
9. James Mauro, *Twilight at the World of Tomorrow: Genius, Madness, Murder, and the 1939 World's Fair on the Brink of War* (New York: Ballantine Books, 2010), 16.
10. Corey KilGannon, "Circling the Unisphere, a Borough's Backyard," *New York Times,* August 26, 2011.

supporters, postwar federal restrictions on cost, location, and occupancy ensured that many projects soon would be less "decent, safe, and sanitary" than the tenement neighborhoods they replaced. And failure to provide adequate relocation opportunities for displaced tenants meant that overcrowding simply moved to the next low-income neighborhood. Once the urban renewal program began in 1949, cities across the nation vowed to minimize public housing in future redevelopment plans and concentrate on providing new homes for more upscale households. The dismal experience with the public housing projects that were actually constructed, and sometimes razed, stigmatized all forms of low-income housing programs, a stigma that endures to the present time.

"Waiting for *Gautreaux*"

The deliberate practice of segregating Chicago's poor black residents in high-rise public housing projects like Robert Taylor Homes was challenged in the federal court system in the epic case of *Gautreaux v. Chicago Housing Authority,* which was filed in 1966 and lasted almost indefinitely. Alexander Polikoff, the lead attorney for the plaintiffs, aptly titled his memoir of the struggle *Waiting for Gautreaux*. Like *Jarndyce v. Jarndyce,* the fictional litigation of Dickens's *Bleak House, Gautreaux* would become a synonym for judicial pettifogging, frustration, and delay.[14]

Dorothy Gautreaux was a young African American CHA public housing tenant and activist (who, sadly, died early in the course of the litigation). On behalf of her and five other low-income black tenants, Polikoff and his team sued the CHA and the U.S. Department of Housing and Urban Development (HUD), alleging that their roles in siting, building, and managing a segregated public housing system violated both the U.S. Constitution, which guarantees all citizens equal protection under law, and the 1964 Civil Rights Act, which bans racial discrimination in programs that receive federal funding.[15] The Polikoff team gathered evidence that black applicants were deliberately steered to the worst CHA projects while low-income whites were eligible for units in white neighborhoods. Expert consultants testified that, "if census tracts were viewed as neighborhoods, . . . of some 21,000 CHA apartments completed or under development since 1949, over 98 percent were in black neighborhoods."[16]

The early stage of the litigation coincided with the freedom marches through white neighborhoods of Chicago and other cities led by the Reverend Martin Luther King Jr. in the late 1960s. The assassination of Reverend King on April 4, 1968, incited riots in many cities and prompted Congress to adopt the long-contested federal Fair Housing Act a week later. Three national commission reports in 1967 and 1968 condemned racism in housing, education, employment, and other facets of American society as a national disgrace and threat to social order.

The *Gautreaux* plaintiffs won an initial victory on July 1, 1969, when Judge Richard B. Austin, as urged by the plaintiffs, ordered the CHA to prepare a plan to construct at least four units of public housing in white neighborhoods for every additional unit built in black neighborhoods, beginning with seven hundred units to be located in the former areas.[17] But entrenched political and neighborhood resistance would drag out enforcement, provoking decades of subsequent legal and political struggle reaching as high as the U.S. Supreme Court and the White House.

Frustrated with CHA's noncompliance, Judge Austin, in September 1971, directed the agency to provide a timetable and to submit a list of locations for new public housing units in white neighborhoods, while rebuking Mayor Richard J. Daley for defying federal law.[18] Subsequently, he ordered the CHA to bypass the city council in the site selection process, a mandate upheld on appeal but still yielding no tangible results. Mayor Daley (with some justification) argued that it was unfair that white city neighborhoods were required to accept CHA units while white suburbs were exempt.

The plaintiffs then proposed a metropolitan-scale plan to scatter public housing among suburbs as well as city neighborhoods. This proposition was rejected by Judge Austin, who did not want to let the city off the hook.[19] He was reversed by the U.S. Court of Appeals for the Seventh Circuit, and the issue was submitted to the U.S. Supreme Court (which, ominously for the *Gautreaux* plaintiffs, had just rejected a metropolitan solution to school desegregation in *Milliken v. Bradley*[20]). Polikoff argued before the court that *Gautreaux* differed from *Milliken* because HUD actively supported segregated housing in the former while the latter case did not involve explicitly unconstitutional action by the defendant school districts. In an 8–0 opinion issued on April 21, 1976, the Supreme Court agreed with Polikoff's concept of a metropolitan-level housing market but did not approve any authority to impose public housing on white suburbs.[21]

The court's acceptance of the concept of a metropolitan housing market reflected the creation and passage of the Section 8 Housing Choice Voucher Program by Congress in 1974. Section 8 rent subsidies theoretically were applicable toward the cost of housing wherever a willing landlord could be found in the city or suburbs. But "willing landlords" were few outside the traditionally black ghettos. Polikoff, despairing of opening up white communities without judicial sanction, wrote later: "The holy grail of a public housing program, in which the middle-class and affluent white neighborhoods of suburbia would have to accept a fair share of the region's public housing poor, was now unattainable."[22] (I discuss the concurrent Mount Laurel, New Jersey, "fair share" housing litigation in chapter 4).

In September 1978, a new judge (Austin died in 1977) approved Polikoff's request for a housing expert to be appointed to review CHA compliance with the *Gautreaux* order. The eminent planning lawyer Richard F. Babcock prepared a detailed report in six months that "exposed the bureaucratic excuses as cant."[23] Even this, however, did not encourage the CHA to comply with the "scattered sites" judicial orders; in 1985, Babcock characterized *Gautreaux* as "Chicago's tragedy."[24] Fortunately, it was a premature epitaph. Although the direct results of *Gautreaux* in Chicago were agonizingly slow to materialize, most of Chicago's infamous high-rise projects were demolished in the early 2000s at the behest of Mayor Richard M. Daley, their sites to be redeveloped as mixed-income, low-rise housing.

More broadly, *Gautreaux* steered national public housing policy away from high-rise, segregated projects and toward scattered, small-scale clusters of housing units that blend in with surrounding areas. The federal Section 8 voucher program established in 1974 and the HOPE VI public housing program launched in 1992 each embodied the spirit of *Gautreaux*. Section 8, which awards federal rent subsidies to low-income families toward the cost of available market units has proved to be a significant alternative to the "hard units" approach of public housing construction. Polikoff regretfully observes, however, that many tenants relocated under both programs remained in largely segregated and often poor neighborhoods, albeit arguably better than the hideous projects which are now history.[25]

From Stuyvesant Town to "Stuytown"

A vast expanse of red-brick apartment buildings that borders the East River in lower Manhattan architecturally resembles the public housing projects of its day, but it had a very different genesis and clientele as a public-private housing development for the white middle class. When they were built, Stuyvesant Town and its sister project Peter Cooper Village comprised the largest apartment development in the United States. The combined project (referred to as Stuyvesant Town, or as rebranded today, "Stuytown") evolved from an agreement made in 1943 between the City of New York and the Metropolitan Life Insurance Company, which was brokered by the ubiquitous Robert Moses. The agreement, as described by the urban historian Martha Biondi, "authorized an unprecedented transfer of state resources to a for-profit private venture, including a twenty-five-year tax exemption estimated at $53 million, the ceding of public streets, and the condemnation of private property, which involved the forced removal of ten thousand residents."[26]

Stuyvesant Town ultimately comprised a complex of 110 eleven-story brick buildings containing 11,250 apartments housing some 25,000 middle-class New Yorkers—a vast supply of affordable rental units near the heart of the city. The eighty-acre project site is sandwiched between the elevated FDR East Side Drive (another Moses project) and First Avenue, extending from Fourteenth to Twenty-third Street. No commercial space, schools, or playgrounds were originally provided inside the project, but shopping is within walking distance. Loosely modeled on Corbusier's Ville Radieuse, the apartment towers are modestly separated by patches of grass and hedges bordered by walkways, with a green "oval" at the center of the community. Stuyvesant Town's high-minded aims were inscribed in 1947 on a bronze plaque (since removed) honoring Metropolitan Life's chairman at the time, Frederick H. Ecker: "Who with the vision of experience and the energy of youth conceived and brought into being this project, and others like it, that families of moderate means might live in health, comfort, and dignity in parklike communities, and that a pattern might be set of private enterprise productively devoted to public service."

Although Stuyvesant Town was not a federal urban renewal project, it anticipated several elements of the 1949 Title I program:

- Designation of a specified urban district as a "slum" or "blighted area."
- Governmental purchase and clearance of the district using public eminent domain powers to compel private owners to sell.
- Resale of the project area to a for-profit redevelopment company at a discount price ("write-down") substantially less than the public costs of acquisition, clearance, tenant removal, and street closures.
- Redevelopment of the site by a corporate land purchaser in accordance with a pre-established master plan.
- Absence of public input in the design or management of the project.

Stuyvesant Town was planned and executed by the city and state (with Robert Moses representing both) and the Metropolitan Life Company. Residents of the area to be razed had no voice.

The 1943 agreement explicitly allowed the company to restrict Stuyvesant Town to white households. Frederick Ecker publicly stated: "Negroes and whites don't mix. If we brought them [blacks] into the development, . . . it would depress all the surrounding property."[27] (That's why the plaque was removed in 2002.) However, unlike at its suburban counterparts Levittown and Park Forest (which I discuss in chapter 4), racial discrimination at Stuyvesant Town stirred up a political and legal firestorm. First, New York's activist mayor Fiorello LaGuardia attempted unsuccessfully to add nondiscriminatory language to the contract with Metropolitan Life. Then a coalition of the American Civil Liberties Union, the NAACP, and the American Jewish Congress filed a lawsuit charging that in subsidizing a racially discriminatory private development, the state and city violated the equal protection clauses of the federal and state constitutions. The New York Court of Appeals in 1949 rejected that claim, upholding the right of Metropolitan Life to rent to whomever it chose regardless of the public subsidies.

The outcome strongly reflected behind-the-scenes legal maneuvering by Robert Moses. Martha Biondi described his impact, as an appointed official, on New York State law: "Moses was the quintessential state activist—he used public authority to literally rebuild the city—yet the government somehow became useless when challenge was promoting social reform."[28] In her opinion, the outrage stirred up by the Stuyvesant Town case helped to launch the fair housing movement, a key element of the civil rights movement of the 1960s.

In addition to its status as the largest apartment complex in the country when it was constructed, Stuyvesant Town has acquired two further distinctions, as the most costly residential real estate transaction in U.S. history and as the nation's largest real estate bankruptcy. The below-market rentals at Stuyvesant Town were mandated as a quid pro quo for the public subsidies extended to the developer by the city. Under city and state rent control laws, many of the project's units have remained to this day well below market rents. On the assumption that most of Stuyvesant Town would soon emerge from rent control, the mega–real estate firm Tishman Speyer, together with various partners, purchased the entire project from Metropolitan Life in 2006 for $5.6 billion.[29] But only three years later, in 2009, Tishman Speyer racked up the largest default ever on its Stuyvesant Town deal, in part owing to a court decision that affirmed the rent control status of many units and required the new owner to reimburse the tenants a total of $200 million.[30]

Today, an Internet search reveals a schizoid picture of Stuyvesant Town. Current management portrays the complex as a hip "Stuytown" set in an "80-acre private park" (as though the buildings were not present), offering sports, festivals, and upscale remodeled apartments at rents upwards of $3,000 per month. By contrast, the Tenants Association reports a grim battle to resist rent hikes that are forcing out long-term residents and a litany of management deficiencies relating to common spaces, noise, trash, laundry rooms, dog litter, and so forth.[31]

A visit to the complex in April 2011 disclosed a demographic schism between elderly residents with walkers—presumably long-time rent control tenants—and a multiracial cohort of young families and urban professionals. The "Oval" is now a busy recreation and community space bordered by an Internet cafe, cinema, bar, and other nouveau enterprises tucked into ground-floor corners of the original apartment buildings. A two-bedroom apartment in Peter Cooper Village, the slightly tonier Siamese twin to Stuyvesant Town, now rents "from $4,400 a month" ($52,800 a year) according to a full-page advertisement in *New York Magazine* in 2011.

Urban Renewal: Panacea or WMD?

The hotly debated Housing Act of 1949 incorporated certain elements of both Stuyvesant Town and the earlier public housing program. Elements borrowed from the latter included: federal funding assistance to local authorities to acquire and clear slum buildings and neighborhoods, with the power of eminent domain as needed, leading to redevelopment of the site for purposes consistent with federal goals and criteria. Title III of the act authorized 800,000 more low-rent public housing units, as proposed by President Harry Truman in his 1949 State of the Union address.

But even the Title III stimulus to public housing paled beside the far more ambitious Title I urban redevelopment program. As William L. Slayton, the federal commissioner of urban renewal, testified in 1963: "Congress clearly indicated that the *clearance and redevelopment of blighted areas was a national objective,* that private enterprise could not do it alone, the public power to assemble land was necessary, and the public costs should be shared by federal and local governments."[32]

Elasticity of Urban Renewal

Urban renewal would prove to be broader, more flexible, and more amenable to local governments and the building industry than public housing with respect to designation of project areas, permissible redevelopment outcomes, and local cost sharing. Although the preamble to the 1949 act referred to clearance of both "slums" and "blighted areas," the latter opened up much broader targets of opportunity for clearance through the urban renewal process. Now, instead of focusing on structures and neighborhoods of abject squalor ("slums"), the term "blighted" could be applied to areas showing signs of age and obsolescence—almost a ubiquitous condition after fifteen years of depression and war. This swept far more urban areas into the maw of the federal bulldozer than the narrower concept of "slum clearance."

Second, the permissible forms of redevelopment would be considerably broadened over time. Initially, the program was confined to "predominantly residential" redevelopment (not necessarily low-income), but the program was amended to allow up to 30 percent of a project to be nonresidential.[33] Thus federal urban renewal subsidies were increasingly available for downtown shopping malls, office clusters, hotels, convention centers, and sports stadiums. The most acclaimed prototype for downtown redevelopment—in fact, completed without urban renewal funds—was Pittsburgh's Gateway Center, a cluster of gleaming stainless steel office buildings, sharing a state park at the Golden Triangle where the Allegheny and Monongahela rivers converge to form the Ohio River.[34] Showcase downtown makeovers subsequently supported by Title I funding included Detroit's Renaissance Center, Hartford's Constitution Plaza, Baltimore's Charles Street Center, Boston's Prudential Center,[35] Cleveland's partially completed Erieview project, and the Gateway Arch District in St. Louis, among many others.

The third area of federal flexibility was the "local cost-share" of project costs. Title I authorized a federal share of two-thirds of the costs of acquiring and clearing project sites, with one-third to be provided by the local urban renewal authority. In practice, the local share could include in-kind routine local expenses that might have occurred without the project, for example, street paving, lighting, landscaping, schools, police and fire stations.[36] (This leniency regarding "local share" would also apply to other federal cost-share programs such as highways and disaster assistance.)

Berman v. Parker

The U.S. Supreme Court unanimously endorsed the expansive interpretation of urban renewal in its landmark 1954 decision in *Berman v. Parker*.[37] The case was brought by local business and homeowners in the Southwest quadrant of Washington, DC, where District and federal authorities planned to demolish an older mixed-race community and replace it with upscale apartments and town houses along with commercial and government office construction. The Southwest urban renewal plan has been scathingly characterized by one historian as "the work of local architects and business leaders . . . [who] were generally opposed to the concept of public housing for the poor because they envisioned an affluent, white metropolitan core." As a result the area "came to be seen as blighted and in need of total redevelopment. That meant the wholesale bulldozing of the quadrant and the removal of its population."[38] Relocation meant forcing people to find alternate housing or move in with relatives in other low-income neighborhoods—such as the isolated ghetto communities beyond the Anacostia River in the Southeast quadrant of Washington, DC—where problems of overcrowding, sanitation, schools, and access to jobs would be worsened.

The plaintiffs in *Berman v. Parker* did not challenge the flagrant racism of the Southwest urban renewal plan.[39] Instead, they posed the narrower question of whether condemnation of private property declared "blighted" would constitute a "public use" as required by the Fifth Amendment to the U.S Constitution.[40] (In fact, the renewal plan envisioned replacing the existing business with another retail use, one presumably more attractive to white shoppers.) Justice William O. Douglas, an ardent conservationist and progressive, wrote the court's unanimous opinion that sweepingly upheld the constitutionality of urban renewal: "The concept of the public welfare is broad and inclusive. . . . The values it represents are spiritual as well as physical, aesthetic as well as monetary. It is within the power of the legislature to determine that the community should be beautiful as well as healthy, spacious as well as clean, well-balanced as well as carefully patrolled. . . . If those who govern the District of Columbia decide that the Nation's Capital should be beautiful as well as sanitary, there is nothing in the Fifth Amendment that stands in the way."[41]

With this imprimatur of the Supreme Court, Title I urban renewal spread rapidly across the nation to cities large and small. By 1965, 772 local governments were participating in the program, over half of them smaller than 25,000, and nearly 1,400 urban renewal projects had been initiated.[42] Although many projects involved some public housing in their redevelopment plans, most residential redevelopment under the program took the form of upscale modernist medium to high-rise apartment buildings, adjoined sometimes by town houses and retail facilities. Some, like I. M. Pei's Silver Towers and adjacent Washington Square Village, both serving New York University faculty and students, are admired expressions of the modernist style: bold, unadorned slabs of pre-stressed concrete grids framing floor-to-ceiling windows, set in a superblock amid parking and greenspaces. Under the direction of Edmund Bacon, Philadelphia's Society Hill redevelopment project likewise blended high- and low-rise apartment buildings with rehabilitation of nearby town houses.[43]

Some projects like Silver Towers and Society Hill have achieved a measure of lasting approval, but many downtown makeovers have been notoriously disappointing. Indoor downtown shopping malls, initially designed and touted by the architect Victor Gruen, have often failed, as in Baltimore, New Haven, Hartford, Springfield (Massachusetts), and San Bernardino, California (to name just a few that I have perversely visited.) New commercial and government buildings stand as statements of the importance of their architects and builders with little relationship to what surrounds them. The architectural critic Ada Louise Huxtable wrote in 1963 with reference to the Arch District in St. Louis: "The arch stands in a curious kind of limbo called urban redevelopment . . . with no relationship to the city at its feet. . . . Downtown St. Louis is a monument to Chamber of Commerce planning and design. It is a businessman's dream of redevelopment come true."[44] In the same vein, the Harvard planner William Alonso observed in 1971: "To save the downtown area, what is needed is a downtown that works well. Many of the new downtown projects are composed mostly of free-standing buildings, handsomely set about in open space, designed as sculpture on a grandiose scale. The emphasis is put on the project as such, not on the downtown area as a part of the urban system."[45]

One poignant description of the sad losses wrought by urban redevelopment is by the journalist Alex Marshall, written about his former home of Norfolk, Virginia:

> By the early 1960s, gone were the century-old burlesque theaters, the old train station, and the fabulous city markets, one built Art Deco–style in the 1930s and the other with medieval turrets in the 1880s. Gone were the often elegant buildings near the water where brokers and other businessmen bargained over the tons of coffee and coal that made their way in and out of the port. Most of all, gone was the tiny network of streets, many of them still cobblestoned, that invoked the memory of the city's oldest days, dating back several centuries. . . .
>
> While tearing stuff down, the city was facilitating the insertion into downtown of giant freeways, many of which were laid atop the oldest neighborhoods. As part of the redevelopment, the old city hall and courts, which formed a central square at the city's heart, were closed, and a giant, windswept plaza with modernist skyscrapers of concrete and glass was built as the new municipal center.[46]

Urban universities such as Yale, Brown, Harvard, Columbia, Penn, and the University of Chicago quickly joined their host cities in adapting urban renewal to enlarge their facilities and reshape their surrounding neighborhoods.[47] New Haven, in collaboration with Yale, rebuilt its downtown in the early 1960s under Mayor Richard Lee and director of urban renewal Edward Logue. The uninspiring results included a new highway connector,[48] some banal apartment buildings, an in-town shopping mall (closed in the 1990s and later converted to luxury apartments), and the New Haven Coliseum and garage complex (both since demolished). More lasting have been Wooster Square and other projects to rehabilitate rather than demolish older New Haven neighborhoods.[49] The University of Chicago vigorously embraced urban renewal to replace blocks of older walk-ups and obsolete storefronts with upscale housing and shopping facilities in its Hyde Park–Kenwood neighborhood.[50] Urban hospitals like Bellevue in Manhattan, Massachusetts General in Boston, and the University of Chicago Medical Center also expanded and redesigned their surroundings through urban renewal.

The campus of the University of Illinois at Chicago (UIC) today stands as an ambiguous monument to the urban renewal era and the passions it aroused. In the late 1950s, a new branch of the state university was proposed to be located somewhere within commuting reach of Chicago's metropolitan population. The site eventually selected, strongly advocated by Mayor Richard J. Daley, lay on the edge of the city's downtown (the Loop) near the intersection of three new expressways and nominally served by public transit. The proposed 105-acre campus would supersede an earlier urban renewal plan already under way to provide new mixed-income housing, and additionally the campus would claim adjoining neighborhoods occupied by Greeks, Italians, African Americans, and Hispanics—the very area served by Jane Addams's Hull House. Local residents under the leadership of Florence Scala petitioned, marched, and litigated to block the university project. The battle was judicially resolved in 1962 in favor of the urban renewal authorities, and the project proceeded to displace eight thousand individuals and 630 businesses, as well as most of the Hull House complex except the original home itself and one other building (thus accomplishing what the 1909 *Plan of Chicago* would have had it been fully implemented).[51]

On a recent midsummer visit, I was struck by the isolation and sterility of the UIC campus. Like the mythical island of Bali Hai in *South Pacific,* Chicago's spectacular skyline looms in the distance beyond the intervening expressways. Like so many relics of urban renewal, the campus was designed as a self-contained modernist ensemble, with no architectural or functional relationship to the exciting downtown that it borders.

The Relocation Quandary

By mid-1967 new dwelling units constructed through urban renewal totaled 41,580—one-tenth the number demolished by then and two-fifths the total number of new houses and apartments eventually completed on urban renewal sites, according to the National Commission on Urban Problems (Douglas Commission).[52] Furthermore, only 10,760 of completed units were in the form of low-income public housing. The commission also found that among urban renewal projects in the pipeline at that time, only 74,000 low- and moderate-income housing units were planned—again, just one-tenth of the total of 760,000 units scheduled for demolition.[53] People evicted from buildings to be demolished obviously were not relocated to new housing created by the program and were largely left to re-house themselves with a meager relocation stipend. Indeed, according to Herbert Gans, "only one-tenth of one percent of all federal expenditures for urban renewal between 1949 and 1964 was spent on relocation of families and individuals, and 2 percent of payments to businesses are included."[54]

In addition to its failure to relocate displaced low-income people of color and small businesses, urban renewal was charged by its early critics with destroying viable ethnic neighborhoods in order to replace them with upscale development. In his 1962 book *The Urban Villagers,* Herbert Gans chronicled the economic and psychological hardships visited on residents of Boston's Italian West End neighborhood, which was redeveloped with what Alexander Garvin would later describe as "a pallid complex of modernistic boxes"—bland apartment towers for the upper middle class.[55] The housing attorney Charles Abrams decried the branding of viable neighborhoods in desirable locations as "blighted" in order to facilitate their redevelopment by private builders. Referring to a planned new apartment development in midtown Manhattan, Abrams wrote: "The site of Kips Bay . . . is a blend of old tenements and stores—but only a ten-minute walk to the United Nations and another five minutes to the Grand Central office complex. Far from being a slum, the Kips Bay renewal site sold to [builder] Zeckendorf at a subsidized bargain price was one which any developer would have bought without a write-down."[56]

Similarly, the Douglas Commission deplored the diversion of urban renewal away from its original mission: "Instead of a grand assault on slums and blight as an integral part of a campaign for 'a decent home and a suitable living environment for every American family,' renewal was and is too often looked upon as a federally financed gimmick to provide relatively cheap land for a miscellany of profitable or prestigious enterprises."[57]

Jane Jacobs famously excoriated the entire canon of urban planning and redevelopment and its foundation on utopian misconceptions of the nature of city and neighborhood: "Cities are an immense laboratory of trial and error, failure and success, in city building and city design. This is the laboratory in which city planning should have been learning and forming and testing its theories. Instead the practitioners and teachers of this discipline (if such it can be called) have ignored the study of success and failure in real life . . . and are guided instead by principles derived from the behavior and appearance of towns, suburbs, fairs, and imaginary dream cities—from anything but cities themselves."[58]

In response to these and other critiques (though none more scathing than Jacobs's), urban renewal morphed into new forms and approaches through subsequent acts of Congress. These turbulent decades produced many new provisions in housing laws—too many and too arcane to cover here, but some high points may be mentioned. The Housing Act of 1954 introduced "rehabilitation" of structures and "neighborhood conservation" as new program elements to mollify objections to total clearance of project areas. The act also required that proposed urban renewal projects conform to a local "workable program" for community improvement, although that concept was never fully elucidated.[59] The Housing Acts of 1959 and 1961 respectively introduced new mortgage subsidy programs for elderly housing and for "low and moderate income" family housing. The latter program, known by the catchy title Section 221(D)(3), for the first time extended federal housing incentives to private for-profit as well as nonprofit builders. It was directed toward helping the large number of households that fell between the maximum income limits for public housing and the minimum levels needed for access to private market housing, with special priority for those displaced by urban renewal. The 1961 act set a goal of 40,000 new apartments per year, but by 1970 a total of only 140,000 units had been completed.[60]

Interstate Highways: "Urban Meat Axe"

The National Interstate and Defense Highways Act of 1956[61] launched the largest public works project in the nation's history. The interstate highway system was originally planned to connect all the nation's major cities with forty-one thousand miles of limited-access new highways at a cost of $50 billion. This immense budget (the entire federal budget at the time was $71 billion and the Marshall Plan to rebuild Europe cost a mere $17 billion)[62] would be funded from taxes on gasoline and other vehicle charges earmarked for the National Highway Trust Fund. The federal share of interstate highway construction and maintenance was an unprecedented 90 percent, with the remainder easily satisfied with in-kind state and local contributions. It was a bonanza for state and city highway departments and the road-building industry. It was the kiss of death for many urban neighborhoods that had somehow escaped the onslaught of Title I urban renewal projects.

The 1956 act was the culmination of decades of growing political influence of the Highway Lobby, a coalition of automobile manufacturers, truckers, oil companies, road builders, and other pro-highway stakeholders. General Motors, Ford, Chrysler, and their competitors devoted huge budgets in the interwar years to promoting new cars as the emblem of the American Dream, along with high-speed highways to drive them on. The culmination of this campaign to promote cars and highways was the GM Futurama and its companion exhibit Democracity—both smash hits of the 1939 New York World's Fair. GM sold its fifty-millionth vehicle in 1954, and car sales by the industry as a whole grew by 37 percent in the following year.[63] The 1956 Interstate Highways Act followed as inevitably as the reelection of President Dwight D. Eisenhower the same year (and for whom the system was later named)

President Eisenhower indeed championed and signed the act based on his admiration for the German autobahn high-speed road network. But he envisioned the interstate system to extend, like the autobahns, from one city to another rather than slashing through the hearts of cities. That, however, was not what city leaders had in mind. In 1953, Robert Moses predictably demanded that the program be available for highways *within* cities, despite the higher costs and disruption that would entail. Moses memorably ranted: "You can draw any kind of picture you like on a clean slate, but when you're operating in an overbuilt metropolis you have to hack your way with a meat axe."[64]

The meat axe proved an apt metaphor. Turned loose to draw new highway corridors as they wished, state and city engineers targeted lower-income neighborhoods to improve mobility in and out of central cities for the more affluent (fig. 3.3). Before the Uniform Relocation Assistance Act of 1970, displaced residents and small businesses were treated even more callously than under urban renewal. And while urban renewal at least had the veneer of social legitimacy as the product of city planning, highways were the province of federal, state, and city highway engineers and their contractors—the mandarins of the Technocrat Decades. Between 1967 and 1970 alone—when Title I clearance had begun to yield to neighborhood conservation and rehabilitation—interstate highways displaced some 168,000 people.[65] Three-quarters of those were in cities, and most, being poor and non-white, were shut out of relocation to white neighborhoods and suburbs.

The adverse impacts of urban highways of course spread far beyond their actual rights-of-way (which often measured three hundred feet or more in width). Title I presupposed that property bordering an urban renewal project would attract new investment and enhance the tax base, but this scarcely applied to urban highways. Of course, the siting and design of cloverleaf interchanges bestowed unearned windfalls on owners of property suddenly in demand for service stations, convenience stores, motels, fast-food restaurants, and the like.

But occupants of property bordering major highways are assaulted by twenty-four-hour traffic noise, diesel fumes, and nighttime lights, and isolated from areas across the highway. Public housing projects like the Robert Taylor Homes in Chicago were frequently built along new interstates because no private investors wanted to locate there and low-income tenants had no other housing choices. Childhood asthma has become endemic in communities bordering major truck corridors, such as Hunts Point in the South Bronx, New York, where one in four children suffer from respiratory illnesses (see chapter 7).[66]

Deterioration of public transportation has frequently accompanied the construction of new highways within cities—the assumption being that anyone not owning a car is irrelevant. As the urban sociologist Robert Bullard has noted: "Not having reliable public transportation can mean the difference between gainful employment and a life of poverty in the ghettos and barrios."[67] More broadly,

Fig. 3.3. "Urban Meat Axe": Chicago's Dan Ryan Expressway.

he argues: "Transportation policies did not emerge in a race- and class-neutral society. Transportation-planning outcomes often reflected the biases of their originators with the losers comprised largely of the poor, powerless, and people of color. . . . White racism shapes transportation and . . . denies many black Americans and other people of color the benefits, freedoms, rewards offered to white Americans. In the end, racist transportation planning can determine where people of color live, work, and play."[68]

Before the 1960s, the New York metropolitan area was interlaced by scenic, noncommercial parkways, many constructed by Robert Moses. But with the infusion of 90 percent federal interstate funding, parkway construction yielded to the creation of heavy truck routes such as the Long Island and Cross Bronx expressways, and the Sheridan "highway to nowhere" that extends slightly over one mile through the heart of the Bronx with no destination. For those communities lying in the path of Moses's highways, his legacy (according to the transportation planner and South Bronx community leader Omar Freilla), "is one of racism and classism, forced removals, splitting of neighborhoods, economic depression, and increasing pollution. Conservative estimates place the number of people forced from their homes for his highways at 250,000—a number that jumps to almost half a million when factoring in the homes bulldozed to make way for 'urban renewal' and other projects. . . . Moses's celebrated highways unleashed forces that gutted stable neighborhoods and sent marginal ones careening over the edge."[69]

Where proposed highways threatened neighborhoods or parks of concern to urbanites with more political clout, they pushed back, as on the Brooklyn Heights waterfront where Moses provided an esplanade above the Brooklyn-Queens Expressway (BQE). (While the esplanade has served mostly the local residents of Brooklyn Heights, the much larger Brooklyn Bridge Park is under construction to provide waterfront recreation to a broader public.)

Manhattan proved to be a hard nut to crack even for Moses. While I-95 crosses the narrow upper part of the island (connecting the George Washington Bridge and New England), no other interstate has survived Manhattan's citizen activists. Most famous was the demise of the Lower Manhattan Expressway: not only was the project canceled, but also Moses was forced to resign in 1968 from his Triborough Authority power base by Governor Nelson Rockefeller.[70] A proposed Midtown Manhattan Expressway similarly was killed by opposition from local business and real estate interests.[71] (I discuss Westway, Manhattan's last proposed interstate, in chapter 8.)

In 1967, plans for the Inner Belt Expressway through Boston, Cambridge, and Brookline encountered a bloc of over five hundred Harvard and M.I.T. faculty, who petitioned the U.S. secretary of transportation to postpone any construction pending development of an integrated transportation plan for the region. Congressman Thomas "Tip" O'Neill famously declared: "The Inner Belt will be a China Wall dislocating 7,000 people just to save someone in New Hampshire twenty minutes on his way to the South Shore."[72] Construction was halted by Governor Francis Sargent in 1970, and I-95 was re-routed around the city on the existing Route 128 beltway. (Boston's post-2000 Big Dig and greenway are also discussed in chapter 8.) Meanwhile, Baltimore residents blocked a proposed segment of Interstate 70 through a middle-income black neighborhood and Gwynns Falls Park. After two decades of controversy, another "highway to nowhere" resulted from the termination in the 1980s of the road at the park boundary and the city line, beyond which it continues westward for two thousand miles without further interruption.[73]

Urban waterfronts bordering harbors, lakes, and rivers also attracted many proposed highways through the 1960s as the decline of maritime activity left a backwash of abandoned piers, warehouses, and seedy neighborhoods in cities across the nation. In some cases, waterfront highway and park development were integrated,

The Crosstown and the Housing Crisis

Adapted from a memorandum of August 9, 1972, that I wrote while working as staff attorney for the Chicago Open Lands Project. The proposed Crosstown Expressway would have slashed through mostly black middle-class neighborhoods to link the city's radial expressways.

Several federal laws clearly declare that no federal-aid highway will be approved for construction *unless and until adequate replacement housing has already been provided for or is built.* Furthermore such housing shall be fair housing, open to all persons regardless of race, color, religion, sex, or national origin. The Chicago Crosstown Expressway . . . will displace 3,470 dwelling units inhabited by approximately 10,400 persons. One-third of these units are single-family dwellings. Almost one-half are owner-occupied. Most are of good structural quality in some of the most recently developed areas of Chicago. Of the families to be displaced, 1,450 have earnings in excess of the ceilings for federally subsidized housing and therefore will be thrown onto the private market.

But the City of Chicago faces a major housing crisis. The Chicago Housing Authority has a waiting list of 8,000 households and 10,000 elderly persons. Demolition ordered by the Housing Court is presently removing over 15,000 apartments of private units from the city's housing inventory, far exceeding the number of new homes and apartments currently under construction. Furthermore, most new units are too expensive for the families to be displaced by the Crosstown, and many are in all-white neighborhoods that are not welcoming to black households. Moving to the suburbs, which are mostly white, also is not a feasible for most potential displacees in the present segregated housing market.

The federal Department of Transportation (DOT) declared in 1970: "It is the policy of DOT that no person shall be displaced by . . . federal and federally-assisted construction projects unless and until adequate replacement housing has already been provided for or is built."[1]

Thus a mere statement that a given number of vacant units may be located somewhere within the City of Chicago would not satisfy this order. Relocation housing must be identified specifically as to location, size, cost, and condition. Furthermore, all housing proposed for relocation must be certified that it is open to persons of any race, color, or religion.

Decent housing in Chicago is a more urgent social need than additional highways. Under the applicable federal policies, the city and state MUST ensure that decent housing of at least the same or better quality is available to displacees before they are forced to move. Since that federal requirement cannot be fulfilled in the existing Chicago housing market, the Crosstown Expressway must be canceled. [It was!]

1. Department of Transportation Order 5620.1, June 24, 1970.

such as New York's Henry Hudson Parkway through Riverside Park, and Chicago's Lake Shore Drive. More often, however, water's-edge highways physically and visually walled off cities from their waterfronts, as did Manhattan's FDR East Side Drive, the BQE and Belt Parkway in Brooklyn, and I-91 through downtown Hartford, Connecticut, and Springfield, Massachusetts. In San Francisco, the Embarcadero Freeway along the Bayfront was partially completed in the 1950s, leading to widespread sentiment to tear it down. In 1989, the Loma Prieta earthquake obligingly damaged the double-decked structure, which was then razed, leading to huge increases in real estate values along the city's newly accessible shoreline.

The New Frontier and Great Society

By the beginning of John F. Kennedy's "Thousand Days" in 1961, public reaction against the technocratic reshaping of urban America through top-down redevelopment and highway programs was gaining momentum. But while urban critics like Jane Jacobs and Charles Abrams expressed their views in books and articles, the victims of racism began to take matters into their own hands. A wave of nonviolent protests against racism began with the lunch counter sit-ins in Greensboro, North Carolina, and elsewhere in 1960. In the following year, the Congress on Racial Equality (CORE) began to organize Freedom Rides to protest segregation in public transportation across the South, inciting violent reprisals from pro-segregationists along their routes. On August 28, 1963, during the March on Washington, Martin Luther King Jr. delivered his immortal "I Have a Dream" speech from the steps of the Lincoln Memorial. At street level, urban riots broke out first in Harlem and Brooklyn in 1964, then in the Watts district of Los Angeles in 1965, Cleveland's Hough neighborhood in 1966, and Detroit in 1967, and in dozens of cities including the nation's capital after Dr. King's assassination on April 4, 1968.

Against this tumultuous background of protests and urban violence, the eight years spanned by the Kennedy and Johnson administrations from January 1961 to January 1969 witnessed an unprecedented and thereafter unequaled national focus on civil rights. Under the complacent Eisenhower administration, inner city blight and economic distress were considered an engineering problem, to be addressed through urban renewal, highway construction, and public housing. With the advent of the Kennedy administration, and especially the personal commitment of Attorney General Robert F. Kennedy, civil rights became a national priority. The new attention to inner cities and civil rights was reinforced by a generation of young activist city mayors during the 1960s, including Jerome Cavanaugh in Detroit, Carl Stokes In Cleveland, Thomas D'Alesandro III in Baltimore, John Lindsay in New York, and Kevin White in Boston.[74]

But the Kennedy administration—terminated by the assassination of the president on November 22, 1963—was too brief, and southern political hostility too intense, to achieve major civil rights legislation.[75] Remarkably, the Kennedy civil rights agenda was embraced by a powerful southern politician who famously hated Robert F. Kennedy: President Lyndon Baines Johnson. In the words of the historian Roger Biles, "The federal government's commitment to America's cities in the post–World War II era reached its apotheosis during the Johnson Presidency"[76] (table 3.1). The former Senate majority leader, now redeemed from the purgatory of the vice presidency, would goad Congress to adopt a host of major federal enactments on civil rights, housing, transportation, and related issues. He was supported and sometimes prodded by such stalwart urban liberals as senators Paul Douglas (Illinois), Edmund Muskie (Maine), Henry Jackson (Washington), Harrison Williams (New Jersey), and William Proxmire (Wisconsin), Representative Henry Reuss (Wisconsin), and mayors Richard J. Daley of Chicago, Robert Wagner Jr. of New York, and Richard J. Lee of New Haven, among many others.

Two major laws, the Civil Rights Act of 1964 and the Voting Rights Act of 1965, were each Kennedy legacies; others were products of Johnson's War on Poverty and Great Society initiatives. The Civil Rights Act, among its many elements, outlawed racial discrimination in public transportation, hotels, restaurants, and similar travel-related enterprises, thus at last legally empowering nonwhites to travel anywhere in the United States. Title VII of the act—the Equal Opportunity Act (EOA)—similarly banned discrimination in hiring, promotion, compensation,

3.1 Major Federal Laws on Civil Rights and Cities Adopted during the Johnson Administration, 1963–1969

1964	Civil Rights Act
	Housing Act
	Urban Mass Transportation Act
	Equal Opportunity Act ("War on Poverty")
	Voting Rights Act
1965	Omnibus Housing Act (Rent Supplements)
	Department of Housing and Urban Development (HUD) Act
	National Capital Transportation Act (Washington DC Metro)
	Highway Beautification Act
1966	Historic Preservation Act
	Department of Transportation Act (DOT)
	Demonstrations Cities ("Model Cities") Act
1967	Air Quality Act
1968	Civil Rights Act (Fair Housing Act)
	Housing and Urban Development Act

and other aspects of public or private employment on the basis of race, color, religion, sex, national origin (age and disabilities were later added as further protected conditions). The Voting Rights Act extended Fifteenth Amendment protection of the right to vote regardless of race to include state and local elections. Beyond these two signature Kennedy legacies, President Johnson, with the support of majorities in both houses of Congress, rewrote the nation's major housing laws in the landmark Housing Acts of 1965 and 1968, and the Fair Housing Act of 1968.

Urban public transportation was another priority for the Johnson administration, which signed into law the Urban Mass Transportation Act of 1964 and the Department of Transportation Act of 1966. Section 4(f) of the latter statute restrained proposed highway construction through public parks and similar environmental sites, "unless 1) there is 'no feasible and prudent alternative' to using the site, and 2) the project includes all possible planning to minimize harm to the site."[77] Along with the Historic Preservation Act of the same year, Section 4(f) anticipated the National Environmental Policy Act of 1970 (signed by Richard M. Nixon) in requiring assessment of the environmental impacts of federally related projects. Such oversight did not, however, extend to residential neighborhoods: targeting low-income communities for new highways and urban redevelopment continued into the 1970s, subject only to the scant benefits accorded to residents and businesses displaced by federal projects under the Uniform Relocation Act of 1970.

This harvest of new federal laws on cities and minorities was urgently crafted by the Johnson administration in the face of relentless Republican and southern opposition, and against a background of growing chaos in the nation's inner cities. A potentially crucial step was the creation of the U.S. Department of Housing and Urban Development (HUD) in 1965 as an umbrella for the housing-related programs already in effect. Also, the Johnson administration made unprecedented use of high-profile blue-ribbon commissions to document the underlying needs of cities and minorities as a foundation for legislative solutions. The National Advisory Commission on Civil Disorders (Kerner Commission), established in 1967, devoted seven months to a detailed study of the pathology of life in urban ghettos. Its report issued on February 28, 1968, excoriated the nation's policies, which were leading toward "two societies, one black, one white—separate and unequal," called for urgent reforms in public policies relating to housing, schools, police, and jobs. Especially controversial was its call for open housing legislation to allow nonwhites to move into exclusionary suburbs.

The Kerner Commission's scathing indictment of racism in urban policies was reinforced by two other blue-ribbon panels: the National Commission on Urban Problems (Douglas Commission) and the Presidential Committee on Urban Housing (Kaiser Commission). In hearings in cities across the country, members of the Douglas Commission: "heard from a steady stream of disgruntled public housing residents, community organizers, civil rights activist. . . . Witnesses further excoriated urban renewal for its wholesale destruction of housing in salvageable neighborhoods and for its wholly inadequate construction of replacement units."[78]

Anti-integration factions in Congress had long and strenuously opposed federal open housing legislation, but the killing of Dr. King finally shocked enough lawmakers to pass the Fair Housing Act (Title VIII of the Civil Rights Act) on April 11, 1968, one week after the tragedy. The act prohibited discrimination in the sale or rental of homes in the private real estate market, striking a significant legal blow to such practices as steering black families away from white communities. Four months later on August 1, the omnibus Housing and Urban Development Act of 1968 was enacted. Among many initiatives, it called for the construction of 26 million new homes over ten years, six million of them for low- and moderate-income households.

The two acts adopted in tandem seemed to be a spectacular breakthrough for civil rights advocates and a turning point in the relationship of the federal government with cities and suburbs. According to Alexander Polikoff, lead attorney in the *Gautreaux* case: "Collectively, the Kerner Report in March [1968], the housing laws of April and August, and the Kaiser and Douglas reports in December laid the rhetorical groundwork for a historic reversal of federal policy toward the black ghetto."[79]

This reversal, of course, was not to be. Growing public outrage against the Vietnam War discouraged President Johnson from seeking another term and doomed Hubert Humphrey's nomination to replace him amid mass demonstrations during the 1968 Democratic National Convention in Chicago. The inauguration of President Richard M. Nixon and Vice President Spiro Agnew abruptly ended the Great Society. Although that era's harvest of new laws remained in place, the determination to apply and enforce them was emphatically rejected by the new Republican administration.

HUD's abortive Model Cities Program, enacted over intense Republican opposition in 1966. was emblematic of the larger narrative of Great Society ideals thwarted by bureaucracy and politics. This program reflected a growing consensus of liberal policy analysts that urban renewal needed to be reconfigured to maintain and revitalize existing neighborhoods rather than destroy and replace them with upscale new development. (This assessment was strongly reaffirmed by all three national commission reports.) Model Cities sought to moderate the insensitivity of top-down one-size-fits-all funding strategies by adapting federal investments to the needs of real urban neighborhoods, as identified, in part, by the neighborhood residents themselves. (The patronizing term "citizen participation" became a mantra for urban planners beginning in the mid-1960s, but too often it meant letting people ventilate for a few minutes at highly structured public meetings with no follow through.) In contrast to the "stovepipe" administration of separate programs for housing, schools, mass transit, job training, human services, and so forth, Model Cities was intended to address all of these needs through grants awarded to cities on a competitive basis. It was the antithesis of conventional Title 1 urban renewal and potentially a valuable adjustment toward more humane urban policies.

Alas, Model Cities foundered in part because HUD could not really let go of the purse strings: mayors of cities awarded grants, such as Richard J. Daley, wanted autonomy and "rejected close supervision by distant Washington bureaucrats."[80] The program was also hampered by inexperience in the art of citizen participation and how to resolve conflicting needs and viewpoints. The program was also faulted by liberals for being too stingy and confusing. After several-dozen applicant cities were awarded planning grants, only nine full-scale Model Cities programs were funded by the end of the Johnson administration.[81] It was an easy target for the anti-urban Nixon administration; essentially it had already self-destructed.

The Renewal Era in Retrospect

The "central city renewal engine" drastically altered American cities physically, economically, socially, and environmentally. Urban renewal, especially its earlier "clearance" phase, left a checkered legacy of public housing projects (many since demolished), failed downtown shopping malls, bland office and condominium towers, and vast empty spaces—acquired, cleared, but never reused. Urban renewal demolished far more affordable dwelling units than it replaced, leaving the inner city black population adrift to migrate from one decaying neighborhood to the next. The Great Society initiatives under Lyndon Johnson provided critical new civil rights protections and redirected federal urban renewal funding into a variety of new housing and neighborhood initiatives such as rent supplements and Model Cities, which generally were casualties of the post–Great Society Republican era under the Nixon, Ford, and Reagan administrations.[82]

American city downtowns and neighborhoods would certainly have lost their 1920s authenticity (romanticized today by New Urbanists) with or without the intervention of federal and state renewal and highway programs. Manufacturing would have migrated away from industrial cities in search of lower wages and taxes, lack of labor unions, and more lenient environmental regulation, no matter how local politicians tried to retain them. Signature departure stores and other downtown retail establishments would likely have succumbed to suburban malls—and more recently the Internet—no matter how much the city attempted to spruce up its center with new parking, indoor shopping complexes, public art, and landscaped plazas. Corporate headquarters would have relocated according to the whims of their executives and the demands of globalization. Older housing that survived the renewal and highway programs would have "trickled down" to lower-income and immigrant populations, continuing the process that long predated the Technocrat Decades. And despite the success of light rail and express bus service in some cities, it is unlikely that the nation would have abandoned its infatuation with cars (and now SUVs) even without the interstate highway system.

Despite all those qualifications, it remains inescapable that the Central City Renewal Engine dictated a uniform template of highway design and commercial renewal that diminished both a city's sense of place and its ability to adapt to changing economics and demographics. Let me make the point through an anecdotal comparison of two Connecticut River cities: Springfield and Northampton, Massachusetts, one heavily impacted by 1960s-era renewal and highway construction, the other relatively unscathed by either program. While comparisons are tricky, I once did a comparative photo walk one Sunday afternoon around the downtowns of both cities.

The center of Springfield—despite a handsome (but no longer used) white Congregational church and City Beautiful–era city hall and symphony hall—is overshadowed by the hulking elevated I-91 that divides the city from the Connecticut River. Urban renewal endowed the city center with a couple of big-name hotels connected by a skywalk, a once-humming but now moribund indoor shopping mall, and a handful of glass and steel office buildings adjoined by small triangles of grass, shrubs, public art installations, and empty benches. Lacking much in the way of eating, drinking, shopping, entertainment, or other people, the downtown was sadly vacant, except for pigeons, police cruisers, and the occasional cell phone gabber (fig. 3.4).

Same afternoon, different city: downtown Northampton swings to the beat of street musicians, conversations (personal and cellular), and the cuckoo walk signal that serenades groups of people crisscrossing the main intersection. Stores and eating places are open, sidewalk cafes are busy, benches are occupied, and people stroll, smoke, sip, yak, and admire one another's dogs and babies (fig. 3.5). Like the people, the buildings of downtown "Noho" are diverse: mostly 3–4 story brick walk-ups with apartments and therapists upstairs and stores and cafes at street level. A former department store houses a bazaar of hip retail and eating establishments. Like misplaced chess pieces, the castellated City Hall and dowager Academy of Music Opera House define one end of Main Street, while a mural-decorated railroad trestle and bike overpass mark the other end. Interstate 91 is well removed from downtown, and the only evidence of "renewal" is the home-grown variety spawned by local entrepreneurs who have converted banks to art galleries, a movie theater to a performance venue.

Of course, many factors account for differences between two places like Springfield and Northampton. The former has high levels of poverty, joblessness, poor schools, and street violence. The latter has a mixed population ranging from homeless to affluent and a blend of business people, professionals, college faculty

Fig. 3.4. The empty Tower Square shopping complex in Springfield, Massachusetts..

Fig. 3.5. Downtown Northampton, Massachusetts, on the same Sunday afternoon as figure 3.4.

and students, blue-collar workers, artists, musicians, and odd-job doers. These demographic and economic contrasts may be both cause and result of the respective strengths and weaknesses of these two riverfront cities. But my point is that, at a very basic level, a city downtown that was relatively untouched by the federal renewal and highway programs has proven much more adaptable, inviting, and resilient than another nearby city that allowed its center to be torn asunder by the top-down, one-size-fits-all template of the Technocrat Decades. Jane Jacobs was absolutely right.

Chapter 4

The Suburban Sprawl Engine

[A suburb is] the city trying to escape the consequences of being a city while still remaining a city. It is urban society trying to eat its cake and keep it, too.
—HARLAND PAUL DOUGLAS, *The Suburban Trend*, 1925

Local governments are not just the small towns of President Nixon's youth or the gold-plated suburbs of his campaign contributors. Local government is the sinking ship of the central city and the makeshift raft of the new suburbs. The privileged of our urban areas have long since fled to their islands of taxing and zoning exclusivity, leaving the less fortunate to flounder.
—RUTHERFORD H. PLATT, unpublished letter to the *New York Times* (excerpt), 1973

"Urban sprawl"—a term as pejorative as "urban renewal"—was a direct result of the second track of America's postwar housing and development strategy: the "suburban sprawl engine." As the reciprocal to the central city renewal engine discussed in the last chapter, this phrase refers to the combined influence of public policies and the private sector on metropolitan growth *outside* of central cities.[1] These contrasting sets of policies jointly reflected what the Kerner Commission on Urban Disorders grimly called in 1968 the reality of "separate and unequal" societies in the United States.[2] A core precept drove both the central city and the suburban "engines": *Cities are where the affluent become more so and the poor subsist, while suburbs are where the American Dream is realized (for those who qualify).*

The postwar suburban sprawl engine was powered by a high-octane blend of government agencies and private business corporations that joined forces to provide millions of young, white middle-class families with new single-family homes in car-dependent subdivisions, and also to make a great deal of money for the corporations in the process. The private sector could never have undertaken such a massive building binge on its own: government at all levels was needed to provide new highways, schools, police and fire stations, water and sewer systems, recreation spaces, and all the other necessities for the upwardly mobile. While lenders, builders, realtors, oil companies, car and appliance makers, and other private-sector stakeholders eagerly embraced the sprawling metropolis, it would not have happened without tax-supported government buy-in. In large measure, suburban sprawl was powered by selective federal subsidies and tax breaks for developers and home buyers, federal funding for highways and other infrastructure, and local zoning codes that shaped how "suburbia" would look and who could live there. The land policy writer Anthony Flint aptly named this unholy alliance: "Sprawl Inc."[3]

Sprawl by Intent

That sprawl has resulted not from neglect or carelessness but from deliberate *intent* has long been recognized by urban researchers such as the economist Anthony Downs, who made the charge in his 1973 book *Opening Up the Suburbs:*

> Urban development in America is frequently described as "chaotic" and "unplanned" because it produces what many critics call "urban sprawl." But economically, politically, and socially, American urban development occurs in a systematic, highly predictable manner. It leads to precisely the results desired by those who dominate it. As a consequence, most urban households with incomes above the national median or somewhat lower enjoy relatively high quality neighborhood environments.
>
> Yet this process is also inherently unjust to millions of other American households. Furthermore, it contributes to rising maladies in many central cities, including high crime rates, vandalism, housing abandonment, low quality schools, and fiscal difficulties. Unfortunately, most Americans do not understand how these ills are partly caused by the very process that has helped them achieve good quality neighborhoods.[4]

A quarter-century after Downs's indictment, Arthur C. Nelson and Thomas W. Sanchez similarly observed a series of priorities in U.S. urban policy favoring "new construction over rehabilitation or reuse of buildings, highways over public transit, conversion of undeveloped land to urban uses over leaving it intact, construction of single-family (owner-occupied) over multifamily (renter) housing, growing areas over depressed ones, and new locations over old ones"[5] Likewise, in 2000, Paul Grogan and Tony Proscio wrote: "Deep anti-urban biases infect much of the behavior of the federal government in both its regulatory regimes and its programs and investments. That historic bias contributed hugely to the decline of the cities in the first place, and it remains a significant barrier to their full recovery."[6]

The suburban sprawl engine labored mightily in the early postwar decades. During the 1950s fifteen million new housing units were constructed, and housing starts would exceed 1.2 million in every year from 1947 to 1964, mostly outside central cities.[7] American central cities grew by ten million while their suburbs gained 85 million between 1950 and 1970.[8] According to the urbanist David Rusk, suburbia tripled in population between 1950 and 2000, whereas central cities collectively grew by only 73 percent over the same period; by the latter year, half of Americans lived in suburbs, as compared with about one-third in 1960 when the nation's population was about equally divided between central cities, suburbs, and everywhere else. But even that modest gain by central cities occurred mostly in the Sun Belt states which are more permissive in allowing cities to annex surrounding areas of new development—"elastic cities," as Rusk coined them[9]—as compared with "inelastic" Rust Belt cities enclosed by a noose of suburbs. San Diego's central city population grew by 75 percent between 1970 and 2000; Phoenix, 126 percent; Las Vegas, 279 percent; and Houston, 58 percent. By comparison, Detroit and Buffalo each shrank by 37 percent and Washington, DC, shrank by nearly one-quarter.[10]

Most metropolitan areas during the heyday of sprawl consumed land much faster than their rate of population growth. Between 1982 and 1997, "urbanized areas" (as defined by the U.S. Census Bureau) increased in area by 47 percent while the nation's population grew by only 17 percent.[11] The Chicago metropolitan area grew by 48 percent in population between 1950 and 1995 while its urbanized land area increased by 165 percent.[12] Overall, the average density of metropolitan America declined from 407 persons per square mile in 1950 to 330 in 2000.[13]

Postwar sprawl initially had its academic champions. The coalescence of overlapping metropolitan areas along America's northeastern seaboard was famously christened "Megalopolis" by the French geographer Jean Gottmann in his 1961 book of that title.[14] Megalopolis was and is a super-metropolitan region extending from north of Boston to south of Washington, DC, sharing a dense network of transportation and communication linkages connecting its urban centers of finance, culture, and politics. Gottmann extolled Megalopolis as "a stupendous monument erected by titanic effort.[15] Furthermore, he declared it to be, "on the average, the richest, best, educated, best housed, and best serviced [urban region] in the world"[16]

This unbridled enthusiasm captured the spirit of the public and private drivers of the suburban sprawl engine. But dissenting voices began to question the process of suburban sprawl and its flip side, urban renewal, as early as the 1950s. Starting with the protests of a few Cassandras—such as Lewis Mumford, William H. Whyte, and Jane Jacobs—the outcry against urban sprawl could be heard far and wide in the planning profession, public media, the arts, and in political discourse from the 1960s to the present. Over time, urban sprawl has been indicted for an ever-wider spectrum of harms, beginning with loss of scenery and productive farmland in the 1950s, and expanding in subsequent decades to include destruction of ecological habitats, increasing damage from floods, air pollution, traffic congestion, social justice, and contribution to climate change and global warming (effects I discuss in chapter 6).

The suburban sprawl engine seemed impossible to stop despite efforts to "manage growth" adopted by planners in the 1970s—and which continue in today's "Smart Growth" era. Like America's military-industrial complex, which continues to wage the Cold War decades after the fall of the Soviet Union, the housing and road construction industries remained dependent on suburban sprawl for their continued prosperity, at least until the housing meltdown that began in 2006.[17]

Anthony Flint has described the "self-perpetuating suburb, marked by a constant cycle of building more roads to get to more homes and more commercial clusters, [that] was established as a permanent feature of the U.S. economy, supported by a framework of government policies. . . . After 1956 [when the Interstate Highway System was authorized] there was no turning back."[18]

In other words, without fundamental change in the underlying public policies driving the process, growth management advocates have been trying to brake a car with the gas pedal stuck to the floor.

Levittown and Park Forest: Contrasting Prototypes

The original purpose of the suburban sprawl engine was to build what the urban landscape historian Dolores Hayden calls "sitcom suburbs"[19]—the fictional small towns of Archie and Jughead, Ozzie and Harriet, and Norman Rockwell's wistful *Saturday Evening Post* covers. (Jackie Gleason's *The Honeymooners* and *The Amos 'n Andy Show* portrayed the urban equivalents, respectively spoofing white and black working-class life in the big city.) Although popular cultural media thus nurtured the myth of small-town suburbia as the apotheosis of the American Dream, the reality was quite different. Far from closely-knit small towns where people put down roots, postwar suburbia would fall somewhere along a spectrum defined by two contrasting prototype suburbs: Levittown on Long Island and Park Forest near Chicago. Both were massive new developments at the edge of their respective metropolitan areas for a largely transient population of young white veterans. But the objectives of the developers of Levittown and Park Forest, at least in creating a sense of place and community, were diametrically opposed.

Suburbanization was well underway before the Great Depression and World War II temporarily suspended new home construction. But pre-war suburbs consisted of individually built homes and tiny subdivisions clustered near streetcar or commuter rail lines—the original "transit oriented development" now revered by apostles of Smart Growth. The postwar era introduced mass production of housing in vast car-dependent, suburban fringe developments (fig. 4.1). The leading pioneer of this new suburban model was William Levitt, the developer of Levittown, a vast expanse of tiny single-family homes covering 1,200 acres of Long Island potato fields, twenty-five miles east of New York City. (He later built two other Levittowns, in Pennsylvania and New Jersey.)

Levitt revolutionized the home construction industry by applying Henry Ford's mass production model to assemble thousands of houses from prefabricated components, using (nonunion) work teams who repeated the same tasks on each site in rapid succession. The Long Island Levittown eventually sold over 17,000 single-family homes at modest cost to white buyers approved for FHA or VA loans (I discuss the effects of government subsidies later in this chapter). Seven houses

Fig. 4.1. The march of sprawl across prime farmland (former farm buildings are at the center).

per acre were constructed on concrete slabs instead of foundations, and a variety of new building materials and machine tools were employed. Each lot fronted on a curvilinear road or cul de sac, a token bow to garden city principles—but minus Ebenezer Howard's gardens and parks. Levitt scrimped ruthlessly on infrastructure and community needs. He installed a cheap cesspool for each house rather than a sewer system. There was no public transportation and no adequate provision for parking of cars. There was no local government, nor even a homeowner association. Primitive roadways within the development were conveyed to the township for improvement and maintenance. Likewise, area school authorities were stuck with the cost of providing schools for Levittown's children.[20]

Nevertheless, Levittown was a huge marketing and financial success. Levitt appeared on the cover of *Time* (July 3, 1950)[21] and enjoyed other media accolades, and his methods were widely imitated around the country.[22] But the uniformity of house design and layout, scarcity of public spaces, and automobile dependency attracted scorn from Lewis Mumford[23] and later from Kenneth Jackson, Dolores Hayden, and other critics of "cookie-cutter suburbia." The economist Edward Glaeser takes such critics to task in his 2011 book *Triumph of the City,* asserting that Levittown "drove highbrow critics like the *New Yorker*'s Lewis Mumford to fits of literary condescension, [but] the town's low prices and relative opulence made it wildly popular with ordinary folks . . . [who] only needed to come up with $400 to buy a home packed with modern appliances and surrounded by leafy space."[24] Only, that is, provided the buyers were white. (Not to be out-condescended, however, Glaeser eighteen pages earlier laments: "It's a pity that too few ordinary people can afford to live in central Paris or Manhattan.")[25]

As Levittown became the prototype for barebones tract development in the late 1940s, its antithesis was evolving at Park Forest, Illinois, on former cornfields thirty miles south of Chicago. Park Forest resulted from a collaboration of the former New Dealer Philip Klutznik and the Chicago developer Nathan Manilow. Their stated intent was to build a viable community, not just a vast subdivision of small tract homes. Unlike Levittown, Park Forest was designed to provide a variety of housing types, beginning with clusters of low-rise rental units, augmented over time by small starter homes and larger single-family homes. Park Forest also provided schools, a community shopping center, playgrounds, athletic fields, and several churches. While Levittown had no local government of its own, Park Forest was established as a self-governing village incorporated under state law. And in contrast to the auto-dependency of Levittown, Park Forest was and is served by commuter rail service to downtown Chicago. In the mid-1950s, the urban sociologist William H. Whyte painted a gently admiring picture of Park Forest as the archetype home of its young, white suburban inhabitants—the "organization man" and his family.[26] "When the doors were thrown open in 1948 the rental courts were islands in a sea of mud, but the young people came streaming out of Chicago. The first wave of colonists was heavy with academic and professional people. . . . Once there, however, they created something over and above the original bargain. Together, they developed a social atmosphere of striking vigor."[27]

Unlike Levittown, Long Island, Park Forest from the outset was a real community that coalesced around its children, its schools, its churches, its shared ambitions, and (according to Whyte) its perpetual state of conflict with the developers. Like Levittown and Stuyvesant Town, it was initially an all-white enclave. Whyte saw Park Forest in the 1950s as relatively "classless." But classlessness stopped "very sharply at the color line," Whyte wrote, remembering several years earlier "an acrid controversy over the possible admission of Negroes. It threatened to be deeply divisive. . . . Though no Negroes ever did move in [by 1956], the damage was done. . . . The sheer fact that one had to talk about it made it impossible to maintain unblemished the ideal of egalitarianism so cherished."[28]

But in 1959, the village established a human rights commission and adopted a fair housing ordinance. According to a documentary film on Park Forest by William Gilmore, Park Forest experienced an orderly transition toward integration in the 1960s. With approximately 25,000 residents as of 2011, Park Forest is about equally balanced by race.[29] (Levittown on Long Island is still about 90 percent white.)[30]

Federal Incentives to Sprawl

Levittown and Park Forest were each initially lily-white, as were most new housing developments across the country. Never mind that black men served honorably in the armed forces (although in segregated units), and black women worked in the newly feminized labor force at home. Pre-war racist attitudes infused postwar housing and real estate practices, often extending to the barring of Jews, Irish, Italians, Asians, and Hispanics and other "undesirables" from suburban or upscale city housing markets. This discrimination flourished with the overt complicity and validation of the federal government through its housing subsidy and financing functions. Federal tax policies and highway subsidies added more fuel to the suburban sprawl engine.

Mortgage Redlining

Before the 1930s, the building and financing of housing was largely a private market function in which the federal government played little role. With the advent of New Deal housing programs, federal agencies assumed a critical role in determining the racial complexion, as well as the economic status, of new suburban communities (where most housing construction took place). The creation of the Home Owners' Loan Corporation (HOLC) in 1933 applied explicit racial and ethnic discrimination in selecting which homeowners and communities should receive federal assistance to stave off mortgage foreclosures. HOLC developed a detailed procedure for assessing the "suitability" of neighborhoods for federal assistance, thus introducing the invidious practice of "red-lining"—mapping patterns of race, ethnicity, and poverty as a guide to the creditworthiness of individual borrowers. In particular, "homogeneity" was adopted by HOLC as a criterion of neighborhood financial security. Charles Abrams criticized the practice in his 1955 diatribe *Forbidden Neighbors:*

> The most effective promoter of class exclusion . . . was the federal government itself. For more than a decade and a half, the United States had a concerted, rationalized, publicized, government-supported program under which a great section of the new generation was set apart, sterilized against infection by alien culture, taught to live and respect only its own kind, trained to oppose intrusion by those who were different. A man should never live with those in a "higher or lower income scale than his own. It is the better part of wisdom to buy in a neighborhood where people are of his own racial or national type," wrote a government housing economist [in 1937].[31]

Kenneth T. Jackson assessed the damage caused by the HOLC not as the result of its own actions, but rather as the outcome of how the commission's appraisal system influenced the financial decisions of other institutions.[32] Compounding this insidious practice, the Federal Housing Administration (FHA) established by the

Housing Act of 1934 would emulate HOLC's appraisal techniques and pro-white suburban bias: "In the [FHA], representing the largest part of the federal housing program, discrimination and segregation were not only practiced but were openly exhorted."[33] Likewise, the Veterans Administration (VA) applied racial and ethnic tests in its postwar housing assistance programs. Between 1946 and 1953, FHA and VA provided or backed loans for ten million new homes nationwide, all subject to redlining and other discriminatory practices.[34]

Racial discrimination in housing of course did not originate with New Deal housing programs. The private market had long engaged in controlling the sale and rental of housing units through racial restrictive covenants inserted in deeds and lease agreements. Thus white homebuyers approved for mortgages by FHA signed private deed restrictions that prohibited them from selling or renting to blacks or members of other disapproved groups.

In 1948, the U.S. Supreme Court in *Shelley v. Kraemer* held that state judicial enforcement of such racial covenants violated the Equal Protection clause of the Fourteenth Amendment: "These are not cases . . . in which the States have merely abstained from action, leaving private individuals free to impose such discriminations as they see fit. Rather, these are cases in which the States have made available to such individuals the full coercive power of government to deny to petitioners, on the grounds of race or color, the enjoyment of property rights in premises which petitioners are willing and financially able to acquire and which the grantors are willing to sell."[35]

The *Shelley* decision was an early victory for the incipient civil rights movement and for Thurgood Marshall, the lead attorney for the plaintiffs.[36] But the decision did not invalidate restrictive covenants per se which were still effective by voluntary consent and by collusion of realtors, lenders, developers, sellers, neighbors, and public officials to enforce racial restrictions without judicial action. (Charles Abrams in 1955 listed fifteen strategies by which covenants were perpetuated through such practices.)[37] Thus by 1960, the eighty-two thousand residents of Levittown, Long Island, were still entirely white despite the *Shelley* decision and the resale of many homes by that time.[38]

Federal Tax Incentives

Owning and occupying a single-family home has long been the sine qua non of the American Dream as a "safe" investment (at least before the post-2006 foreclosure crisis) and an incentive to good community citizenship. Renting a home, especially in multi-unit buildings, was long associated with the "old urban neighborhood" and its immigrant and low-income tenants. Upward mobility within the white, organization-man mainstream of the first three postwar decades was calibrated by successive moves to ever-larger and more costly owned homes in the "right" communities. While single-family home ownership accounted for less than half of the nation's housing stock during the pre-war decades, it jumped from 43.6 percent in 1940 to 61.9 percent by 1960, and has hovered at about two-thirds of all housing units ever since.[39] Within metropolitan America, the homeowner-ship rate in suburbs has been much higher than in central cities. Between 1960 and 2000, the homeownership rate ranged from 70 to 73 percent in suburban areas, but from 47 to 51 percent in central cities.[40]

The federal income tax code provides three major deductions for home own-ership costs not available to renters: interest on mortgage loans on principal or second homes, property taxes paid to local governments, and "points" paid to lenders at the time of purchase. In 1999, Tom Daniels, a planner at the University of Pennsylvania, calculated that the buyer of a $150,000 home with a $120,000, 30-year mortgage at 8 percent, and who has a gross income of $60,000, would save about $3,900 in federal tax annually during the early years of the loan.[41]

Mortgage subsidies by some estimates cost federal taxpayers over $100 billion per year.[42] While two-thirds of taxpayers (owners and renters) use the "standard deduction" and thus do not benefit from these home ownership write-offs, the more affluent do very well indeed. Roger Lowenstein's *New York Times* report in early 2006 explains the deduction from its 1913 origins and the impacts of ever-increasing home and mortgage size: "The rewards are greatly skewed in favor of the moderately to the conspicuously rich. On a million-dollar mortgage . . . the tax benefit is worth approximately $21,000 a year." Lowenstein compares that to median home figures in 2003—$1,680 for the buyer in the 25-percent tax bracket with a $140,000 mortgage—but is quick point out that "a little over half of the benefit is taken by just 12 percent of taxpayers, or those with incomes of $100,000 or more."[43] The federal tax deduction for property taxes paid to local jurisdictions further subsidizes home ownership.

Not only are renters who itemize their deductions deprived of such tax benefits, but they also miss out on the opportunity to accumulate equity in a rising housing market and "trade up" to a larger home, or to cash in the equity bonus for college tuition or riotous living. In other words, the separate tracks of federal housing policy for cities and suburbs for decades determined who would or would not be entitled to gain wealth through home ownership. This inequity has blurred over time as home ownership in cities has become more common through cooperatives and condominiums, and more affluent people of color are increasingly able finan-cially and legally to purchase homes in suburbs. The poor of any race, however, remain mired in the financial dead end of renting, hoping at least for public rent subsidies when needed. And the housing market downturn and foreclosure crises dating from about 2006 have shattered the dreams and solvency of many African Americans and other minority first-time home buyers.

In 1954, at the height of "sprawl fever," Congress provided another huge incentive to urban sprawl (without racial implications however) in the form of "accelerated depreciation" of commercial property. This allowed developers to depreciate (write off) the capital cost of building shopping malls, hotels, big-box stores, and offices in seven years, rather than over the expected lifetime of the development of, say, thirty years (although many malls don't last that long). Dolores Hayden observes that this provided a "gigantic hidden subsidy for the developers of cheap new commercial buildings."[44] It also encourages shoddy construction, poor maintenance, and rapid turnover of commercial properties. Accelerated depreciation stimulates construction or purchase of commercial buildings as tax shelters intended to show a loss and offset taxable income from other sources.

Sprawl is indirectly promoted by yet another federal tax deduction, namely for interest earned from state and municipal bonds. This tax break allows states, local governments, and special districts to borrow funds in the bond markets at lower interest rates than would be required if such payments were federally taxable. The provision of course applies to bonds issued by central cities as well as to suburbs and rural jurisdictions. But during the go-go years of postwar sprawl, suburban governments and their special district "alter egos" financed construction of roads, sewer and water systems, schools, parks, and other infrastructure through a combination of bonded indebtedness (subsidized by the federal tax code) and direct state and federal grants (funded by taxpayers). Thus, federal tax policies encouraged duplicating urban infrastructure in suburbia—at higher cost per household served due to lower densities—rather than devoting such resources to maintaining and enhancing facilities already available in the cities.[45]

In addition to accelerated depreciation and municipal bond subsidies, the promoters of sprawl invented other tax incentives to help fill up new outlying office and industrial parks and shopping malls, often at the expense of central cities from which many such businesses relocated (fig. 4.2). The business and development strategist Jeb Brugmann pointed out the ramifications for Chicago: "In a brilliant strategic move against the city, suburban municipalities developed new forms of market governance. Municipal land grants, tax holidays, and site development incentives traded off today's revenues and land values against a long-term future stream of industrial and commercial taxes. Faced with the choice between a subsidized new industrial landscape and full-cost capital investment in established city locations, Chicago's manufacturers moved to the suburbs. . . . By 1965, more than half of the manufacturing jobs in metropolitan Chicago had shifted to suburban highway corridors. Between 1965 and 2000, Chicago lost 70 percent of its manufacturing jobs."[46]

Similarly, nearly all job creation in metropolitan Atlanta during the 1990s occurred outside the central city. The northern suburbs added 272,000 jobs, about 78 percent of the region's total between 1990 and 1997, and another 20 percent of jobs located in the south suburbs. Only 4,500 jobs—1.3 percent of the regional total—actually were created in the city of Atlanta.[47] Since, the suburbs were (and

Fig. 4.2. "Sprawl as Kudzu": commercial development facilitated by tax policies.

still are to a lesser extent) predominantly white and upscale, this required inner city black and Hispanic residents to endure costly and time-consuming commutes to reach suburban jobs.

The same competition between central city and suburb for economic growth and tax base plays out at smaller geographic scales, as with a major IBM laboratory next door to, but not within, the city of Poughkeepsie, New York, in the Hudson River valley. The city's downtown in the 1940s, as Harvey K. Flad and Clyde Griffen describe in *Main Street to Main Frames,* was a prosperous commercial center with three department stores anchoring a traditional Main Street. IBM, however, chose to build its facilities in the still-rural town of Poughkeepsie (which surrounds the city on three sides), avoiding the political, racial, and ethnic complexities of the latter. The *town* thus received a huge economic boost while the *city* struggled to save its Main Street economy, even as it housed and serviced most of the region's poor and black population.[48] This is a powerful example, among thousands of others, of the ironclad rigidity of local political boundaries that became established sometime in the past for reasons long forgotten, but endure as Chinese walls between separate units of local governance, inexorably shaping the checkerboard of municipal winners and losers.

Interstate Highways: Back to the Futurama

The federal Interstate Highway System of course contributed immensely to suburban sprawl as well as inner city devastation. The migration of the white middle class and the businesses that employed and served them depended not only on tax incentives but also on free (or cheap, in the case of toll roads) highways connecting suburban fringe development with everywhere else. The primary arteries of the Interstate Highway System span the nation from coast to coast and border to border, with perimeter beltways and myriad connecting spurs slashing through built-up areas. With the federal Highway Trust Fund providing 90 percent of land acquisition and construction costs, the 46,000-mile interstate system redrew the geography of America in terms of housing, jobs, land use, politics, and lifestyle. Stephen B. Goddard described interstate highways as "the cathedrals of the car culture, and their social implications were staggering. Within a decade, they would alter beyond recognition where and how Americans lived, worked, played, shopped, and even loved. Watershed changes loomed in politics, agriculture, and land economics."[49]

Interstate highways simultaneously enabled and encouraged development of previously remote rural land at or beyond the metropolitan fringe, for example in Levittown and Park Forest. As the radius of a metropolitan area lengthens, the total supply of land within reach of development expands geometrically, resulting in progressively patchier, lower-density, and less-coordinated patterns of land use with greater distance from the built-up core. ("Leapfrog development" is a favorite term of the anti-sprawl literature.)

Older commuter rail suburbs like Hastings-on-Hudson (New York), Montclair (New Jersey), or Downers Grove (Illinois)—graced with tree-lined streets, comfortable homes, and good schools—evolved during the postwar decades from middle-class, white-collar enclaves to more expensive sanctuaries for the professional and corporate elite who still required, and could afford, propinquity to the central city. Further out, rank-and-file "organization men" and their families gravitated to lower-cost raw subdivisions on the urban fringe, too widely scattered to support extension of commuter railroads and, unlike Park Forest, seldom located next to an existing rail line. Driving to work, to play, to shop, to socialize, to recreate became the American "new normal": Broadacre City achieved.[50]

Unlike highways through built-up urban neighborhoods, which provoked mass displacement and civic outrage, the interstate system unfolded with less rancor in the open countryside, in part because fewer people lay in its path. And rather than fighting new highways, rural landowners often pulled every political string to be traversed by a new highway corridor, and thus enriched by generous purchase payments. Even greater windfalls were realized by owners of land near future interchanges where gas stations, truck stops, chain restaurants, and overnight lodging facilities would cluster.

Major beltways such as I-495 around Greater Boston and its counterparts elsewhere became lined with postmodern glass office buildings for high tech and financial giants, converting sleepy rural towns into instant Silicon Valleys. Trading their slower pace of life for the fast lane of global capitalism, communities bordering new highways gained huge tax windfalls with relatively few new tax burdens, since the corporate workforce would generally live in other communities. (Sharing of tax base enhancement between windfall communities and their less-fortunate neighbors was mandated for the seven-county Twin Cities metro area by the Minnesota Fiscal Disparities Act of 1971, but the experiment has not been replicated elsewhere.)[51]

The very largest of these highway-dependent complexes—typically located at mega-nodes where two or more interstates cross, or near airports or suburban public transit stations—were dubbed "edge cities" by journalist Joel Garreau.[52] Tysons Corner, Virginia—one of Garreau's archetypal edge cities—mushroomed from a rural crossroads into an economic colossus due to its proximity to the I-495 Capital Beltway surrounding Washington, DC, as well as to I-66 and the Dulles Toll Road. Dolores Hayden described Tysons Corner in 2003: "Inside a knot of freeways and arterials sit unrelated high-rise and low-rise buildings, a vast assemblage of houses, apartments, garages, shopping malls, fast-food franchises, and corporate headquarters."[53]

Tysons Corner and the immense complex surrounding Dulles International Airport (another federally funded growth magnet) in 2011 was semantically upgraded to an "aerotropolis" by the business consultant John D. Kasarda: "The Dulles Toll Road was, in effect, the template for the high-tech avenues I would see later in Dubai, Doha, Guangzhou, and even Dallas."[54] Kasarda claims that such complexes represent the future of global urbanization: "Airports and their aerotropoli are supplanting downtowns and competing edge cities as the nexus of white-collar work throughout the world."[55]

But Garreau found that edge cities—as spin-offs of highways, airports, and rapid transit stations—defy conventional models of governance, community, and sense of place and lack soul.[56] Regarding governance, he described a typical New Jersey edge city named after a highway interchange ("287 and 78") as having "no overall leader, no political boundaries that define the place. It is governed only by a patchwork of zoning boards and planning boards and county boards, and townships boards . . . swirling like gnats—not any elected ruling structure."[57] The same would apply, presumably, to aerotropoli, which essentially are edge cities on steroids. Described as hermetically sealed, totally private realms of corporate power by Kasarda and his co-author Greg Lindsay, aerotropoli offer a rather frightening model for "the way we'll live next," a phrase that serves as an apt subtitle for their book.

Paving the Farms: The Illinois Tollway Extension Battle

Not all rural farm owners are delighted to be in the path of a limited access highway. In the early 1970s, the Illinois State Tollway Authority proposed to extend its existing East-West Tollway from Aurora (on the western edge of metropolitan Chicago) about 85 miles farther to the small farming towns of Sterling and Rock Falls, Illinois. A free interstate highway (I-80) already paralleled the route of the proposed new highway, calling into question how a toll road serving a handful of small towns would be economically justified. Furthermore, certain property owners in the proposed highway corridor opposed the splitting up of their farms, and the prospect of new development (aka "sprawl") eagerly predicted by the Tollway Authority. In support of a lawsuit filed by affected farm owners, I wrote a column for the Chicago Tribune, *from which the following is an excerpt:*

Illinois rural residents have traditionally been the least vocal regarding projects of an environmentally deleterious nature. The philosophy that "any change must be progress, and progress must be for the better" has long prevailed in the farm regions. However, with increased plundering of rural areas for urban wealth, country folks are starting to wake up.

The disruptive effects of the extension are maximized by its diagonal route, which cuts all property lines at irregular angles. While 2,565 acres are actually required for the right of way, another 5,463 acres are severed or isolated from the farms to which they belong. This not only drives up the cost of the project but, more important, vastly complicates the business of farming. Altogether 214 farms are severed in the process.

The total outstanding indebtedness of the Toll Highway Authority . . . could have been paid off in nine years. . . . Instead the authority in 1970 acted to expand and prolong its indebtedness (and its existence) by issuing $135 million in new bonds to finance construction of the extension. This decision, in which toll road users in [metropolitan Chicago] had no voice, committed such users to pay for the extension out of their own pockets until the year 2020.

To devote the entire proceeds of the Illinois Toll Highway System to the construction of unneeded and unwanted additional tollway mileage such as the east-west extension is truly "one of the greatest absurdities that has by governmental action come to be believed" [quoting Bertrand Russell].[1]

As eventually built, I-88 proved to be an engineering fiasco because of wet soil conditions that caused the highway pavement to ripple giving a sensation of riding a low-grade roller coaster. On my one trip on the road to DeKalb, Illinois, around 2005, little traffic was observed.

1. Rutherford H. Platt, "Paving over the Farms: The Road Must Go On," *Chicago Tribune*, April 9, 1972.

Zoning Gamers: The Struggle to Open the Suburbs

With great irony, land use zoning—the capstone achievement of the progressive city planning movement in the 1920s—would become the primary instrument of suburban rejection of demographic and economic diversity. Zoning in the backslapping world of local politics became a parody of the high expectations of its early supporters. Recall the planner and lawyer Alfred Bettman's stirring argument in 1926 to the U.S. Supreme Court in *Village of Euclid v. Amber Realty Co.* (as quoted in chapter 2): "Zoning is based upon a thorough and comprehensive study of developments of modern American cities, with full consideration of economic factors of municipal growth, as well as the social factors. . . . The zone plan is one consistent whole, with parts adjusted to each other, carefully worked out on the basis of actual facts and tendencies, including actual economic factors, so as to secure development of all the territory within the city in such a way as to promote the public health, safety, convenience, order, and general welfare."[58]

Four decades later, another zoning authority, Richard F. Babcock, painted a very different picture in his influential 1966 book, *The Zoning Game:* "The running, ugly sore of zoning is the total failure of this system of law to develop a code of administrative ethics. Stripped of all planning jargon, zoning administration is exposed as a process under which multitudes of isolated social and political units engage in highly emotional altercations over the use of land, most of which are settled by crude tribal adaptations of medieval trial by fire, and a few of which are concluded by confessed ad hoc injunctions of bewildered courts."[59]

What went wrong? That question has preoccupied generations of legal and planning scholars. The short answer is "money, race, and localism." The high-minded "citywide" perspective of early champions like Alfred Bettman and Edward Bassett withered as zoning authority was atomized among myriad cities and suburbs. After its jump-start in New York City in 1916 and its approval by the U.S. Supreme Court in 1926, zoning was embraced by every major U.S. city (aside from the famous exception of Houston, Texas), as well as thousands of counties and smaller municipalities.

Babcock's *The Zoning Game* among many other zoning critiques of the era stimulated a cadre of progressive land use lawyers and planners (whom I will call "zoning gamers") to try to remedy the excesses of zoning while keeping its familiar framework intact. The primary forms of zoning abuse in the eyes of its critics can be summarized as follows:

- *Exclusionary zoning*—the use of zoning to restrict or prohibit apartments or smaller homes affordable by families of modest means, (especially members of racial or other minorities).
- *Fiscal zoning*—the use of zoning to minimize local property taxes by encouraging revenue-generating activities such as shopping centers and industrial parks while discouraging revenue-demanding uses such as lower-cost homes for families with children (regardless of race).

- *NIMBY (not in my backyard)-ism*—the adoption of zoning and other legal means to resist the location of unwanted uses, facilities, or activities within the municipality (e.g., regional incinerators or toxic waste disposal sites, prisons, mental health facilities, oil refineries, halfway houses, and drug clinics).

Two widely utilized exclusionary zoning provisions were (and are) oversized minimum-lot requirements, and apartment bans and other restrictions on multi-family projects. These "low-hanging fruit" were in fact opposed not only by social progressives but also by some land developers who sought greater profits through higher density than allowed under local zoning codes. In the first of two landmark decisions in suits filed by developers, the Pennsylvania Supreme Court in 1965 rejected a four-acre minimum lot imposed by a Philadelphia suburb with a resounding lecture on regionalism: "Zoning is a tool in the hands of governmental bodies which enables them to more effectively meet the demands of evolving and growing communities. It must not and cannot be used by those officials as an instrument by which they may shirk their responsibilities. Zoning is a means by which a governmental body can plan for the future. . . . Zoning provisions may not be used . . . to avoid the increased responsibilities and economic burdens which time and natural growth invariably bring."[60]

Five years later, that statement was repeated by the same court in throwing out a total ban on apartments in a suburban community challenged by another developer.[61] A portentous footnote to that opinion stated, "as long as we allow zoning to be done community by community, it is intolerable to allow one municipality (or many municipalities) to close its doors at the expense of surrounding communities and the central city."[62]

Meanwhile, "zoning gamers" at Rutgers University led by the law professor Norman Williams determined that of 474,000 acres of vacant buildable land in four New Jersey counties, 99 percent was zoned for single-family homes with apartments excluded. Minimum lots of one acre or more were required for 77 percent of buildable land.[63] In the landmark case of *Southern Burlington County NAACP v. Township of Mount Laurel*,[64] such empirical evidence convinced the New Jersey Supreme Court to decide unanimously that Mount Laurel "has acted affirmatively to control development and to attract a selective type of growth . . . and that through its zoning ordinances has exhibited economic discrimination in that the poor have been deprived of adequate housing, . . . There cannot be the slightest doubt that the reason for this course of conduct has been to keep down local taxes on *property* . . . and that the policy was carried out without regard for . . . *people,* either within or without its boundaries."[65]

Audaciously, the court ordered "developing communities" in New Jersey to revise their zoning laws to accept a "fair share" of the regional demand for lower-cost housing. But those terms and the issues around them remained unclear. Meanwhile, lawsuits piled up. In 1983 the New Jersey court issued a second opinion[66] that reaffirmed its earlier mandate and provided detailed criteria to guide local government compliance. Mayors across the state were outraged, and Governor Thomas H. Kean called the *Mount Laurel* opinions "communistic."[67] But he nevertheless signed a state Fair Housing Act in 1985 creating the Council on Affordable Housing.

By 2001, over three hundred suburban jurisdictions had submitted affordable housing plans, but only about forty thousand low- or moderate-cost units had been constructed or renovated, a meager result from a quarter-century of zoning reform efforts in New Jersey. Over time, the *Mount Laurel* strategy foundered at the hands of anti-growth advocates who portrayed the state program as a mandate for housing sprawl.[68] And unlike *Euclid,* which fanned a wildfire of zoning in the 1920s, New Jersey's attempt to rein in exclusionary zoning had little influence elsewhere.

Since his election in 2009, New Jersey's governor Chris Christie has sought to eliminate the state Council on Affordable Housing and dismantle the entire *Mount Laurel* process. The editors of the *New York Times* on January 28, 2013, called for reaffirmation of the *Mount Laurel* doctrine in two cases pending before the New Jersey Supreme Court.[69]

Zoning reformers in Massachusetts pursued a different strategy with the adoption of the 1969 Zoning Appeals Act (aka the "Anti-Snob Zoning Act").[70] The act authorized affordable housing developers (such as public housing authorities or nonprofits like Habitat for Humanity) to request a "comprehensive permit" from local zoning authorities waiving density or multiple-unit zoning restrictions. If the locality denies such a permit, or attaches "uneconomic conditions," the developer may request a state housing appeals committee to override the local government and permit the development if the committee determines that the proposed project would be "consistent with local needs."[71]

As of 2001, the Massachusetts Anti-Snob Zoning Act had made reasonable progress in opening up communities to affordable housing. Of 633 housing applications reviewed, local zoning authorities denied 29 percent, granted 17 percent, and approved 54 percent with negotiated conditions. At the state level, the housing appeals committee upheld the local community in 13 percent of appeals, overruled it in 25 percent, and approved negotiated settlements in 38 percent of cases, with the rest withdrawn or dismissed. The housing researcher Sharon Krefetz of Clark University concluded that the act "has created small toeholds, but the walls of suburban exclusion remain high."[72]

Apartments, subsidized housing, and group homes have become more widely distributed in cities and suburbs since the 1960s. In part, this reflects rising demand from singles, childless couples, and "empty nesters" rather than constitutional attacks on exclusionary zoning. But at least, "affordable housing" has finally become a respectable part of the canon of Smart Growth and New Urbanism, as I discuss in chapter 7. The quixotic efforts to reform zoning in the 1960s and 1970s may be considered a "sacrifice fly" that advanced other affordable-housing "base runners" in later decades.

Chapter 5

Battling the Bulldozer: The Indiana Dunes and Other Sacred Places

> *The Indiana Dunes are to the Midwest what the Grand Canyon is to Arizona, and Yosemite to California. . . . They constitute a signature of time and eternity: once lost the loss would be irrevocable.*
>
> —CARL SANDBURG, letter to Senator Paul H. Douglas, June, 17, 1958

The Indiana Dunes: Drawing a Line in the Sand

One balmy afternoon in early September 1967, a group of new graduate students in the University of Chicago Department of Geography, including me, perched on a sand dune facing Lake Michigan to discuss humans and the Earth. Our interlocutor was Gilbert F. White, an international authority on water resources and natural hazards. "Mr. White" (University of Chicago faculty traditionally eschew the title of "Doctor" as superfluous) posed a series of Socratic questions, weighing our replies with his kindly but skeptical gaze. Little did we then appreciate that the restless and windblown landscape that surrounded us was the "birthplace of ecology," an object of a six-decade (by then) preservation battle, and an epic line drawn in the sand on behalf of people, place, and nature in the path of development.

The site of our afternoon colloquy was a steep-sided "blow-out" or hollow formed by wind-blown sand facing across a narrow beach to the vast sparkling lake that extended three hundred miles northward. We sat amid one of the last remnants of the presettlement landscape of sand ridges and hillocks, glacial moraines, wetlands, and scrub forests that once stretched for many miles along the lakeshore from the present site of Chicago across the northern edge of Indiana into Michigan. In 1967, most of that landscape had been eradicated by steel mills, refineries, power plants, and rail and highway corridors. The dunes landscape still surviving was disturbed here and there by a handful of beach cottage communities. We were hunkered down in the last major oasis of undisturbed "living dunes" in northern Indiana, other than a state park a few miles to the east.

Then as now, the spot where we sat was a world apart from its industrial surroundings: a soft, quiet panorama of undulating dunes sculpted by wind and rain and storm waves (fig. 5.1). The Indiana Dunes region (also known as the Dunelands) is much more complex, geologically and biologically, than a single row of eroding sand hills or glacial bluffs like those on the outer beach of Cape Cod. Over millennia, gradual fluctuations of the lake level created successive ridges of dunes,

Fig. 5.1. Beach and foredune facing Lake Michigan at the Indiana Dunes. Courtesy of
J. Ronald Engel.

leaving swamp and ponds in between, in all extending up to several miles inland
from the present shoreline. Moving inland, the dune ridges are progressively older,
more diminished by weathering, and more heavily vegetated. Just behind the massive and denuded foredunes, patches of beach grasses, dusty miller, barberry, and
other hardy species take hold in pockets sheltered from the wind. A few steps
further inland (and earlier in time), accumulating deposits of soil nurture woody
shrubs like beach plum and viburnum. Beyond the shrub zone, worn-down dunes
and kettle holes dating back to the last glaciation sustain oak and pine forests on
higher ground and bog habitats in the hollows. The naturalist Edwin Way Teale,
who spent, boyhood summers there, described his beloved dunes as "a strange,
tormented battleground where the wind and the root were ever at war—the wind
striving to move the sand along, the vegetation seeking to anchor it down. . . . It
was the mystery of the far-away and the wildness of the dunes which stirred the
imagination."[1]

Although obscure perhaps to East and West Coast cosmopolites who only look
seaward, the battle for the Indiana Dunes was much more than a regional conservation dispute. The extraordinary physical nature of the Dunes region—and
the complexity of social, cultural, economic, and political interests which clashed

over its fate for decades—placed the Indiana Dunes in the pantheon of American environmentalism, in the words of Carl Sandburg as "a signature in time and eternity."

A Geographical Vortex

Much of the Dunes mystique and scientific importance stems from the region's unique centrality in the Midwest and the nation. Think of it as a geographic vortex or hub where ecology, economy, ideology, and culture swirl and interact: this heterogeneity generated not only a wealth of biodiversity, but also a legacy shaped by a century of social and cultural innovation as well as political machinations. This would, in turn, portend the broader emergence over succeeding decades of new perspectives concerning the balance of economic growth versus people and nature in our expanding metropolitan milieus.

Geographically, the Indiana Dunes lies at the crossroads of East and Midwest, as well as North and South. With its distinctive jumble of industry and nature briefly glimpsed from a plane headed into Midway Airport from the east, the Dunelands marks the gateway to the largest metropolis of the Midwest. The region also straddles a divide between southward-looking Indiana (politically and culturally) and the more "northern" states of Illinois and Michigan.

The Dunelands has long been famous as an *ecological* vortex or "mixing zone" between cool and warm climate zones, and between the wetter deciduous forests to the east and the drier prairie province to the west.[2] As documented by many naturalists—Henry C. Cowles, Donald Culross Peattie, Teale, and others—plants and wildlife representative of all those regions are found intermingled by the complexity of microclimates of the Indiana Dunes. In 1899, the University of Chicago botanist Henry C. Cowles published his early research conducted at the Dunes, which would establish him as the "father of the science of ecology":

> There are few places on our continent where so many species of plants are found in so small a compass. . . . Within a stone's throw of almost any spot one may find plants of the desert and plants of rich woodlands, plants of the pine woods and plants of swamps, plants of oak woods and plants of the prairies. Species of the most diverse natural regions are piled here together in such abundance as to make the region a natural botanic preserve. . . . Here one may find the prickly pear cactus of the southwestern desert hobnobbing with the bear-berry of the arctic and alpine regions. . . . Nowhere perhaps in the entire world of plants does the struggle for life take on such dramatic and spectacular phases as in the dunes.[3]

Sadly for its biological riches, the Dunes region has long been an important *economic* vortex for steel production and other heavy industry. Iron ore comes by huge lake vessels from northern Minnesota; coal arrives by rail from southern Indiana and Illinois. Water, electrical power, and a large labor force are close at hand. In 1908, U.S. Steel established its immense mill complex and new city at Gary, Indiana, after clearing the sand dunes from the area. Other steel mills, oil refineries, and power plants soon followed, as did highways and an industrial harbor. A dense network of railroads and highways connected the Dunes region to the rest of the country, and the St. Lawrence Seaway provided shipping access to Europe and beyond. By World War II, the shoreline of Lake Michigan between Gary and Chicago had become, in the words of the urban geographer Harold M. Mayer, "one of the world's largest concentrations of heavy basic industry and the largest steel-producing agglomeration in the world."[4] East of Gary, however, the shoreline remained undisturbed for the time being.

The Dunes also is a *cultural* vortex whose beaches, ever-shifting dunes, and shaded inland hollows have long attracted recreationists, naturalists, and summer residents. With the opening of the South Shore Railroad in 1908, the Dunes became a summer mecca for those whom J. Ronald Engel, in *Sacred Sands: The Struggle for Community in the Indiana Dunes,* called "the creative spirits of the Chicago renaissance of the Progressive era,"[5] beckoning, among others, the landscape architect Jens Jensen, sculptor Loredo Taft, architect Louis Sullivan, poets Harriet Monroe and Carl Sandburg, artists Frank V. Dudley and Earl H. Reed Sr., geologist Thomas Chamberlain, educator and philosopher John Dewey, and of course, Henry Cowles. Hull House founders Jane Addams and Ellen Gates Starr were early Dunes advocates, and Settlement House influence, wrote Engel, would "be readily apparent in the leadership of the Dunes movement through the 1970s."[6]

Challenging Big Steel

Organized support for preserving the Dunes first coalesced around the Prairie Club of Chicago, which conducted walking tours through the Dunelands beginning about 1911. In 1916, the progressive Chicago businessman Stephen Mather, as the first director of the National Park Service, called for creation of a Dunes national park along twenty-five miles of remaining natural shoreline. That proposal was sidetracked by the outbreak of war, but the Prairie Club successfully lobbied for the creation of the Indiana Dunes State Park in 1925, protecting 3.3 miles of shoreline and 2,200 acres of dunes hinterland.[7]

The fate of the remaining undisturbed shoreline between the state park to the east and the steel plants to the west remained unresolved.[8] Midwest Steel purchased 750 acres of pristine dunes in 1929 for future development. In the 1950s, Bethlehem Steel, the nation's second-largest steel company, acquired 4,300 acres of the central core of the Dunelands through a series of insider deals involving politicians, railroads, and land investors.[9] Bethlehem, Midwest, and their allies called on Congress to fund a harbor to allow construction of a fully integrated steel complex. William Peeples, reporting for the *Atlantic Monthly* in 1963 on the history of pressure politics affecting the Dunes, expressed what had seemed in the 1950s like impending doom: "All the stops were pulled out. Indiana would issue revenue bonds to pay for its share of the $70 million port project, and Congress would appropriate at least $25 million in federal funds. Nothing, it seemed, could now save the dunes."[10]

To challenge this industrial/political juggernaut, local residents and their allies in the Chicago region formed the Save the Dunes Council in 1952. Picking up where the Prairie Club had left off in the 1920s, the council quickly sought to arouse public support through guided walks, lectures, a documentary film, art exhibits, fliers, media contacts, and political advocacy. For the complacent, business-centered 1950s, this provoked an epic standoff between social progressivism and pro-industry conservatives.

By the late 1950s, the conflict crystallized around whether the federal government should fund a harbor to serve the steel industry or alternatively purchase the remaining dunes for a national park. Playing David to the industrial Goliath, the Save the Dunes Council appealed in desperation to one of the U.S. senators from Illinois, Paul H. Douglas, the progressive University of Chicago economist with long ties to the Dunes. Senator Douglas's first Dunes bill, filed in 1958, requested Congress to acquire and preserve 3,600 acres, including nearly five miles of unspoiled beach and dunes as a national monument (fig. 5.2). To supporters gathered on the threatened site, Senator Douglas cast the Dunes issue in national terms as "a symbol of the crisis that faces all Americans. It is as though we were standing on the last acre and were faced with a decision of how it should be used. . . . In essence, it foreshadows the time not far removed when we shall in all truth be standing on that last, unused, unprotected acre and be wondering which way to go."[11] Save the Dunes eventually assembled 250,000 signatures on petitions supporting the bill. The *New York Times, Washington Post,* and *Louisville Courier-Journal,* as well as most of the Chicago press, supported the national park editorially.[12]

The Douglas proposal, however, was unprecedented in at least four respects. First, national parks had previously been carved out of lands already in federal ownership, whereas the Indiana Dunes had to be acquired by purchase or gift from private owners, primarily Bethlehem and Midwest Steel (over their strenuous objections). Second, unlike the "crown jewels" of the National Park Service such as Yosemite, Grand Canyon, and Yellowstone, the Indiana Dunes was located within a vast metropolitan region and would serve a predominantly urban population within daytrip range. Third, the Dunes park was proposed explicitly to protect fragile ecological habitats, in potential conflict with its other main purpose of public recreation. Fourth, it was unheard of for a senator to sponsor legislation to create a national park in another state. Douglas was pilloried by the Indiana political establishment for carpetbagging. Undaunted, Douglas would introduce a total of nine bills, with various co-sponsors, between 1958 and 1966.[13]

The destruction of the Central Dunes began in earnest in 1959 when Midwest Steel began building a $100 million "finishing plant" on its property without

Fig. 5.2. Senator Paul H. Douglas (center) with members of the Save the Dunes Council, 1958. "Save the Dunes" (special bulletin), May 26, 1958. Courtesy of the Save the Dunes Council.

waiting for a decision on the port. In 1962, the conflict became more ugly and more urgent as Bethlehem Steel began to bulldoze the rest of the Central Dunes (which it owned) and—adding insult to injury—selling the sand to Northwestern University for use as landfill for campus expansion. Even though the port had not yet been approved, Bethlehem Steel, with encouragement from Indiana politicians, thus willfully eradicated the very heart of the Dunelands, clearly to render the national park challenge moot.[14]

Paul Douglas and Save the Dunes did not surrender. With the Central Dunes now relegated to art, poetry, science articles, and memory, the next park bill in 1963 dropped that focal area from further consideration and requested instead a mélange of beachfront and inland parcels, collectively totaling about 9,000 acres. By now, the park no longer posed a threat to the port complex; both Indiana senators, Birch Bayh and Vance Hartke (both liberal Democrats), joined Paul Douglas in co-sponsoring the new park bill in tandem with parallel legislation to authorize the industrial harbor.[15] In 1966, Congress approved both projects: the Indiana Dunes National Lakeshore and the Burns Harbor industrial complex would become ill-matched future neighbors.[16]

The National Lakeshore legislation authorized $28 million to acquire some 8,100 acres of shorefront and inland ecological sites, greatly adding to the 2,200 acres already protected in the state park.[17] But this only marked the beginning of a new and lengthy phase of the Save the Dunes battle: a separate act was still need to appropriate funds for the park, which was vigorously opposed by pro-industry politicians like Indiana congressman Charles Halleck. Among a procession of threats to the dunes were proposals for a nearby nuclear power plant, a major airport, and a huge railroad-marshalling yard. While those were never built, harmful effluents from the steel mills, a coal-fired power plant, and truck and rail operations have impaired the recreational and ecological functions of the park.[18]

Senator Douglas lost his reelection bid later in 1966, depriving the Dunes advocates of their political "white knight." But with the help of other moderates in Congress from various states, Save the Dunes continued to plead for federal funds to get the park underway and to add further sites of ecological or recreational importance to its authorized area. The park was formally dedicated on September 3, 1972, but many years of further delays, setbacks, and small victories were required to realize much, if not all, of the goals of the Dunes movement since the early 1900s.[19]

The Meaning of the Dunes

Although the loss of the Central Dunes to the steel industry was a bitter disappointment to Dunes advocates, the 1966 compromise at least established a starting point for later enlargement. In 1972, Paul Douglas noted what Congressional approval of the Indiana Dunes National Lakeshore had reaped: "It was a decimated park, to be sure, that we had obtained [in the 1966 act]. . . . But with all the weakness of the new bill, the way was opened for a bigger park. . . . With determination on the part of Congress and the nation, we would have a 12,000-acre park within twenty or thirty years."[20] That prophecy would indeed be realized: the 8,200-acre National Lakeshore authorized by Congress in 1966 has been enlarged many times, now totaling 15,000 acres scattered among many units intermingled with industrial facilities, second-home communities, a power plant, junkyards, and other urban detritus (fig. 5.3).

Despite the fragmented state of the National Lakeshore, the National Park Service has wisely administered it with the ongoing oversight of the Save the Dunes Council. Embedded within a metropolitan area of nine million people, the park today hosts over two million visitors annually in search of the time-honored Dunes pastimes—recreation, education, relaxation, and renewal. The Lakeshore's science staff nurture and interpret "the highest plant species diversity per acre of any park in the National Park system"[21] The Paul H. Douglas Center for Environmental Education[22] provides environmental education to park visitors and the nonprofit Dunes Learning Center offers residential programs in Dunes ecology to thousands of students and teachers annually.[23]

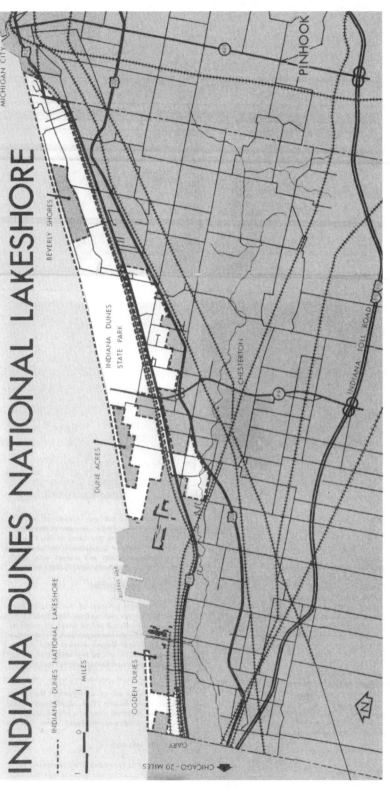

Fig. 5.3. Map of the IDNL saved through the efforts of three generations of Dunes advocates.

Beyond the actual park itself, the Dunes struggle represented an epic exercise in social and cultural democracy, according to Engel, the resident historian and philosopher of the Dunes:

> The campaign to save the Dunes emerged as part of an insurgent movement in the Midwest to reform the democratic faith of the nation. Based largely in Chicago, and rooted in the ethos and politics of Progressivism, this movement sought to sever the identification of democracy with competitive individualism. It conceived the meaning of democracy to be equal freedom in community, or the "cooperative commonwealth." For some members of the movement, the authentic vision of community inherent within the democratic experience was larger than human community alone. They yoked the revolutionary ideals of freedom, equality, and fraternity to the ecological principles of unity and interdependence among all forms of being.[24]

In a sense, the Indiana Dunes harked back to the urban populism that inspired and informed the design of Olmsted and Vaux's Central Park, which would, as Frederick Law Olmsted himself described it, "supply the hundreds of thousands of tired workers, who have no opportunity spend their summers in the country, a specimen of God's handiwork that shall be to them, inexpensively, what a month or two in the White Mountains or the Adirondacks is, at great cost, to those in easier circumstances."[25] But unlike the signature Olmsted park and its counterparts, the Dunes was not a "pseudo-rural countryside" created within a city through engineering and landscape design.[26] To the contrary, the Dunelands was in itself a large-scale swathe of "God's handiwork" that lay in the path of urban destruction. The Dunes struggle, which sought to protect the land for the benefit of people, place, and nature, marked a "prophetic chapter in the history of American public life."[27]

Environmentalism and Urbanism: A Slow-Blooming Alliance

It took some time, however, for the prophetic chapter of the Dunes experience to be replicated in other settings around the country. The establishment of the Indiana Dunes National Lakeshore in 1966 attracted far less national acclaim than its predecessor, the Cape Cod National Seashore, five years earlier. The latter, while immensely valuable, resulted from the Kennedy family's personal influence and a cooperative state and Congress, with relatively little opposition. Saving the Dunes, by contrast, was an uphill battle waged over two (now three) generations against the arrayed power of the steel industry, railroads, power companies, and the Indiana political establishment. It was in fact one of the nation's first democratic triumphs over the technocracy of economic development. Yet it did not even earn a mention in two histories of American environmentalism: Robert Gottlieb's *Forcing the Spring* (1993) and Adam Rome's *The Bulldozer in the Countryside* (2001).

A trite explanation for the lack of interest in the Dunes experience could have been that the movers and shakers of the new environmentalism in New York, Boston, Washington, San Francisco, and Seattle—and even Chicago itself—paid little heed to a dispute over sand dunes in northern Indiana. It simply lacked the charisma of Cape Cod, Storm King on the Hudson, the Adirondacks, the Everglades, Glen Canyon in Arizona, or old growth forest in Oregon and Washington.

Another explanation might be that the Dunes battle was essentially the byproduct of Chicago urban progressivism, dating back to Jane Addams and her disciples in the settlement house movement. While the Dunes pilgrims sought to preserve the Dunes as the birthplace of ecology, they also sought to make the complex accessible to ordinary people, not just the scientific and affluent cognoscenti, a goal fulfilled in the park's hosting of two million visitors annually.

This populist vision contrasted with the nonurban—even anti-urban—ethos of the budding environmentalism of the 1960s. Two of the movement's holy writs, *Sand County Almanac* (1949) and *Silent Spring* (1962), were written respectively by the U.S. Forest Service wildlife manager Aldo Leopold and the marine biologist and nature writer Rachel Carson. Both represented the new genre of articulate field scientist–philosopher more at home in the raw outdoors than city neighborhoods like Chicago's Hyde Park (although Henry Cowles must have been comfortable in both worlds). Similarly, Henry Beston's *The Outermost House* (1925), Marjory Stoneman Douglas's *Everglades: Rivers of Grass* (1947), Edward Abbey's *Desert Solitaire* (1968), John and Mildred Teal's *Life and Death of the Salt Marsh* (1969), and John McPhee's *The Pine Barrens* (1968) and *Encounters with the Archdruid* (1971) drew their inspiration not from urban populism of the 1920s but from Emerson and Thoreau, from Hudson River School artists such as Thomas Cole and Frederick Church, and from George Perkins Marsh's *Man and Nature* (1864). John Muir's immortal plea, to "Dam Hetch Hetchy! As well dam for water tanks the people's cathedrals and churches,"[28] perfectly expressed the Thoreauvian view that wilderness should trump society's practical needs (in the case of Hetch Hetchy Canyon, a water supply for San Francisco).[29]

The disconnect between environmentalists and cities in the United States reflected the rootedness of environmentalism in the conservation movement of the early twentieth century, the era of Theodore Roosevelt, Gifford Pinchot, and Horace Albright. Such influential organizations as the Sierra Club, Izaak Walton League, National Wildlife Federation, Audubon Society, and Nature Conservancy brought to the environmental table in the 1960s and 1970s a long tradition, instilled by their founders and board members, of primary concern with wildlife and their habitats, natural resources, streams and wetlands, and protection of wilderness.[30]

The schism was reinforced by the persistent disinterest in urban environments among leading natural scientists. At the 1955 symposium titled "Man's Role in Changing the Face of the Earth"—a seminal event at the dawn of environmentalism—

the ever-prescient but solo voice of Lewis Mumford warned that the modern city tends to "loosen the bonds that connect [its] inhabitants with nature and to transform, eliminate, or replace its earth-bound aspects, covering the natural site with an artificial environment that enhances the dominance of man and encourages an illusion of complete independence from nature."[31]

But cities would receive short shrift from environmental scientists for decades. The 1965 Conservation Foundation book *Future Environments of North America*[32] entirely ignored urban areas even though they would be the future environments of most North Americans. The 1988 National Academy of Sciences volume titled *Biodiversity*, edited by the Harvard biologist Edward O. Wilson, devoted only seven out of 520 pages to "urban biodiversity." Another landmark survey of environmental change, edited by B. L. Turner and colleagues, *The Earth as Transformed by Human Action* (1990), also gave short shrift to urban areas although they are among the areas most "transformed by human action." In 1992, Al Gore's *Earth in the Balance: Ecology and the Human Spirit* entirely avoided the uncomfortable truths of ecology and the human spirit in cities. Finally in 2003, an essay by William E. Rees called "Understanding Urban Ecosystems: An Ecological Economics Perspective" sought to bridge what Rees deplored as "the legacy of the Enlightenment in western culture . . . that sees the human enterprise as somehow separate from and above the natural world. Such humanity/nature apartheid is even evident in most of our academic disciplines."[33]

The natural adversaries of the old guard conservation organizations have been the industries that exploited natural resources: mining, forestry, hydropower, and tourism, as well as their government agency enablers like the Army Corps of Engineers, the Federal Power Commission, the U.S. Forest Service, and the Bureau of Reclamation. The view of nature as "out there" (i.e., in exotic places accessible only to scientists and affluent ecotourists) has often been reinforced by well-meaning natural history museums, botanical gardens, zoos, and aquaria, many of whose board members were interchangeable with those of the mainstream conservation organizations. Robert Gottlieb describes further how, "through much of the late nineteenth and twentieth centuries, mainstream environmental groups functioned as male-dominated organizations, both in terms of leadership and conceptual framework. The early protectionist and conservationist approaches derived largely from the nostalgia regarding the passing of the frontier and . . . [the rise of] urban and industrial forces. . . . These traditional interpretations of the roots of environmentalism contain powerful gender implications."[34] Gottlieb neglected to extend the discussion to include racial implications (but might well have done so).

The absence of working people, minorities, and women (in the earlier years at least) from the inner sanctums of mainstream environmental organizations long insulated their board members and administrators from the threats to ordinary people and communities from, for example, toxic wastes, brownfields, urban stream degradation, and environmental injustice. Tellingly, the Love Canal toxic waste disaster of the late 1970s in the industrial city of Niagara Falls, New York, came to national attention not because of a major environmental organization but through a local housewife, Lois Gibbs, and the Love Canal Neighborhood Organization. Their advocacy in the best tradition of Jane Addams led to the adoption of the federal Comprehensive Environmental Response, Compensation, and Liability Act of 1980 (aka the Superfund Act) and many state counterparts.[35] Similarly during the 1980s, local housewives focused public attention on health crises stemming from industrial water contamination in Woburn, Massachusetts (the subject of the book and film, A Civil Action),[36] and Toms River, New Jersey.[37]

Old guard conservationists believed, paternalistically, that protecting wilderness and scenic wonders per se would benefit the larger society—the poor included—regardless of the cost and time required to visit such remote and exotic natural areas Protected natural areas were justified, in the words of the historian Robin W. Winks, as "useful, or uplifting, indeed ennobling, to mankind."[38] For some, this stance persisted up to the present century. The New York Times columnist Thomas Friedman quoted an executive of The Nature Conservancy as admitting, "We spent the 20th century protecting nature from people, and we will spend the 21st century protecting nature for people."[39]

The Last Landscapers: New Frontier for the Old Guard

The social upheavals of the 1960s were legion and legendary: civil rights, feminism, anti-war, anti-nuclear, and anti-corporate. In contrast to the youthful and demonstrative members of those causes, a more polite revolution began to unfold during the decade among members of the suburban "chattering classes" who reacted against the loss of farmland, scenic vistas, stream valleys, salt marshes, patches of native vegetation, and historical or cultural sites to the suburban sprawl engine. The perceptive young journalist William H. Whyte first deplored this dreary process in his 1957 essay called "Urban Sprawl."[40]

Over the next decade, fueled by a rash of media coverage and think-tank reports, the "loss of open space" in metropolitan America became a cause célèbre for affluent white suburbanites. This was the "City Beautiful" generation of the early 1900s reincarnated in the era of space travel, Yogi Berra, and the Beatles. But in place of gardens, fountains and monuments, the dinner conversations of this era centered on land trusts, scenic easements, zoning amendments, and the other paraphernalia of open land preservation popularized in books like Charles Little's Challenge of the Land[41] and William H. Whyte's The Last Landscape.[42] In the spirit of the latter work, this new generation may be aptly nicknamed "last landscapers." In general, these were good citizens with open minds and checkbooks who helped to protect many natural, agricultural, cultural, and architectural resources from the bulldozer.

Women played central roles in many if not most open space preservation efforts of the 1960s and 1970s through garden clubs, the League of Women Voters, and individually. The environmental historian Adam Rome observed that in "the domestic sphere—unlike the world of politics and business—women did not have to wait for men to lead they way."[43] Hundreds of open space struggles went undocumented, but Bernice Popelka has provided a memoir of the saving of the five-acre Peacock Prairie about thirty miles northwest of downtown Chicago. Popelka describes herself and her friends as "suburban women of the 1960s, with husbands and families. We had free time to become active stewards of our communities."[44]

A chance conversation in 1965 drew Popelka's attention to a scrap of unplowed native prairie a mile from her home: "The variety of vegetation quickly entranced me. I saw an uncommon beauty here. Tall grasses, some as high as six feet tall, gently waved in the breeze. . . . Stunning blue flowers peeked out from among the grasses. Various asters and goldenrods dotted the field with purple, gold and white. . . . Butterflies and bees hovered over this splendor."[45] (One is reminded of Edwin Way Teale's rhapsodic memories of the Dunes.) Bernice Popelka devoted the next three years to seeking scientific, political, and financial support for preserving what would eventually become the "James Woodworth Prairie." (Also see the case study of Thorn Creek Woods in the box below.)

This genteel revolution of course had mixed motives. On the one hand, open space served as a respectable "sheep's cloak" to conceal the "wolf" of exclusionary zoning. Opening the suburbs to affordable housing, minorities, and the poor (as sought in the *Gautreaux* and *Mount Laurel* legal battles described in chapters 3 and 4) was not their cup of tea. Automobile-centered, single-family suburbia worked just fine for them and their friends, except there was becoming too much of it. The real problem for them was not to dismantle the suburban growth engine but to manage it and direct it away from their "back yards."

On the other hand, the 1960s open space movement helped to redirect the focus of traditional conservationists from the "ennobling landscapes" of exotic and distant places to the "last landscapes" of farms, meadows, swamps, and woods next door. As with the quest to save the Indiana Dunes, urban open space preservation assumed quasi-spiritual significance for many of its partisans. Like the Save the Dunes members in the 1950s, citizen open space advocates learned to evaluate the scientific attributes of a threatened site, to arouse public interest through walks, petitions, and media coverage, and to influence the political process. Armed with petitions, maps, photographs, and philosophy, enlightened amateurs could sometimes alter the seemingly inexorable course toward development. Planners' dogmas, attorneys' maxims, and city fathers' economics all may wither before the arguments of the aroused and informed private citizen.[46]

Saving Thorn Creek Woods

A personal postscript: In the early 1970s, as staff attorney for the Chicago Open Lands Project, I played a role in helping to preserve Thorn Creek Woods (TCW), a 900-acre hardwood and riparian forest in Will County, just across the border (and a street) from Park Forest, Illinois. TCW was threatened with decimation for a "new community" proposed by Lewis Manilow, the son of Park Forest's co-developer Nathan Manilow, who hoped to expand on his father's achievement with another landmark development. The new community of Park Forest South would blend residential, shopping, parks, schools, and even a new state university—a foretaste of Smart Growth ideals today. But the proposed treatment of the forest was not smart. Early designs for Park Forest South proposed housing throughout TCW with parking lots colored green!

Thorn Creek Woods was saved through one of the nation's first applications of the National Environmental Policy Act (NEPA), signed by President Nixon on January 1, 1970. NEPA requires an "environmental impact statement" to be prepared for proposed "major Federal actions significantly affecting the quality of the human environment." The major federal action in this case was the authorization of federal loan guarantees to Park Forest South by the U.S. Department of Housing and Urban Development (HUD) under the New Communities Program of the 1968 Housing Act.

At the urging of Open Lands and local supporters of Thorn Creek Woods, the new U.S. Council on Environmental Quality (CEQ) sent an official from Washington, Bill Matuszeski, to walk through the woods with a group of us. He agreed that the project required a NEPA environmental assessment to be prepared by HUD. The developer, faced with a possible lawsuit to enjoin the federal subsidy, agreed to cooperate with federal, state, and local authorities in preserving TCW, and to contribute some of the land if the rest was purchased. (We called it the Jigsaw Plan.) Today Thorn Creek Woods is a regional forest reservation and nature education facility. The novelty of our strategy was featured in a front page story in the *Wall Street Journal* (October 27, 1970) that read in part: " 'We're trying to use the Council [CEQ] as an ombudsman,' says Rutherford Platt, a Chicago attorney representing the Open Lands Project, one of the forest preservation groups. 'HUD would like to smooth over our objections. But the Council has a unique position under [NEPA] to bring them out into the open.' "

One of the nation's foremost "citizen advocates" during this period was Laurance S. Rockefeller, dubbed "Mr. Conservation" by his biographer Robin W. Winks, in whose assessment Rockefeller did "more than any other living American to place outdoor issues . . . clearly on the table."[47] Unlike his brothers who pursued corporate and political careers, "LSR" combined real estate development of ecotourist resorts with a passionate commitment to protecting natural areas and public park creation. In 1958, President Eisenhower appointed him to chair the Outdoor Recreation Resources Review Commission (ORRRC) established by Congress that year.[48]

Without the benefit of computers or Internet, ORRRC staff and consultants assessed both the demand for, and available supply of, outdoor recreation facilities for a broad range of activities at various scales. Its overview report[49] with twenty-seven background reports (totaling 4,400 pages) was delivered to President Kennedy in 1962. Kennedy's signature park, the Cape Cod National Seashore established in 1961, was a first step in a new era of urban-oriented national parks established under the Kennedy and Johnson administrations, including the Indiana Dunes National Lakeshore. The National Trails and National Scenic and Wild Rivers acts, both adopted in 1968, were also legacies of the ORRRC study.

The report did not radically depart from the conventional thinking of the day and sounds rather banal to a present-day reader; it listed "driving for pleasure" as the nation's favorite outdoor recreation activity in 1960, as illustrated with photographs of white middle-class families in tail-finned cars motoring to picnic sites in "the country." No mention was made of the still-widespread racism in public accommodations, which prevented nonwhites from traveling freely and using public parks equally with whites (segregation in hotels and restaurants was finally outlawed, legally if not always in fact, by the U.S. Supreme Court in 1964.)[50]

ORRRC nevertheless fostered a broader awareness of the need for public open spaces of various sizes, functions, and locations. It made two key recommendations that were adopted. First, the commission proposed that a federal Bureau of Outdoor Recreation (BOR) be established to coordinate federal and state recreation activities and actively promote new recreation initiatives. This was swiftly put into effect by administrative order of Secretary of the Interior Stewart Udall, followed by Congressional ratification in 1963.[51] BOR would play an active role in recreation planning and funding until abolished during the Reagan Administration. The second legacy of the ORRRC study was the creation of the Land and Water Conservation Fund (LWCF) by Congress in 1965.[52] With financial revenues earmarked from offshore oil and gas leases and other sources, the fund supported federal land acquisition as well as state and local open space programs through matching grants provided by the state or local project sponsor. By 1993, the fund had provided over $3 billion, matched by state, local, and private sector money, and had added 3 million acres to the federal public domain for recreational purposes. Laurance Rockefeller would later declare ORRRC to be "one of the most successful commissions in history in terms of legislative results."[53] The LWCF matching grants in turn helped to stimulate billions of dollars in state and local bond issues for open space acquisition and management into the new century.

Reinforcing the growing open space movement, in suburbs at least, landscape design practitioners were hit like a tsunami by Ian L. McHarg's 1968 book *Design with Nature.*[54] Blending his trademark Scottish bombast and innovative computer graphics, McHarg argued that land development must "work with" nature rather than try to overwhelm it. He pioneered the use of multilayer analysis of geographical factors affecting project sites, such as steep slopes, wetlands, aquifer recharge areas, forests, prime farmland, floodplains, historical and cultural features, and scenic beauty. William H. Whyte in *The Last Landscape* applauded McHarg's message as "the gospel of 'physiographic determinism,' which roughly translated means nature ought to come first."[55]

The gospel according to McHarg was most immediately applied to plans for affluent suburban settings such as the Brandywine Valley near Philadelphia and the "valleys" northwest of Baltimore. But McHarg also devoted a chapter to Staten Island, New York, and another to the use of spatial mapping of public health, crime, and other social indicators in Philadelphia. He inspired a new generation of landscape designers to uncover hidden urban waterways and restore patches of nature in city neighborhoods. Anne Whiston Spirn, a McHarg student and disciple, enlarged upon her mentor's evangelism in her 1984 book *The Granite Garden.*[56]

The Indiana Dunes crusade anticipated by several decades Spirn's admonition that "nature in the city must be cultivated, like a garden." But the Dunes advocates would add that "cultivating" implies nurturing not just the biodiversity in the city but also the interaction of people with nature and with each other. Fortunately, by the late 1980s, people, place, and nature were becoming recognized as critical elements of urban sustainability and habitability. This was the segue to the Humane Decades.

Chapter 6

Legacies of Sprawl: A Witch's Brew

> *The ultimate effect of the suburban escape in our time is, ironically, a low-grade uniform environment from which escape is impossible.*
> —LEWIS MUMFORD, *The City in History*, 1961

> *All Americans pay for sprawl with increased health and safety risks, worsening air and water pollution, urban decline, disappearing farmland and wildlife habitat, racial polarization, city/suburban disparities in public education, lack of affordable housing, and the erosion of community.*
> —ROBERT D. BULLARD, *Sprawl City*, 2000

What a strange and dysfunctional metropolitan America we have created. Five decades of efforts to "manage growth" have amounted to the equivalent of "whistling in the wind" against the suburban sprawl engine driven by government and corporate technocracy. Sprawl has continued to flourish like kudzu, paving and building over farmland, forests, desert, wetlands, prairie, mountainsides, barrier islands (see figs. 4.1, 4.2). Disparate swatches of yesterday's sprawl, scattered among a myriad local government fiefdoms, share little except their physical linkages via labyrinthine highway networks on which tens of millions of drivers of cars and trucks waste time and money struggling from one place to another.

The United States has tragically squandered its wealth and talent in failing to provide a better model of metropolitan growth for itself and the developing world. We won World War II, reached the Moon, and outlasted the Soviet Union—but our own people cannot easily travel between home and work (assuming they have both), or fulfill their daily needs with reasonable speed, comfort, and economy. Furthermore, the miseries of the current recession—especially home foreclosures and unemployment—relate in part to the massive miscalculation of builders, lenders, and their government enablers regarding the type, design, and location of new development over the recent past.[1] Today we are faced with the worst of both possible worlds: on the one hand, local governments and the private market reject centralized planning control to achieve more rational land use patterns; on the other hand, local government fiefdoms through their zoning and tax strategies, and social preferences, distort the potential efficiencies of a purely private market, such as the "streetcar suburbs" of the 1870s to 1920s beloved by New Urbanists.

The ugly or monotonous appearance of sprawl (alluded to in Mumford's epigraph above) has aroused decades of outrage from urbanist critics. William H. Whyte in the late 1950s wrote in his essay "Urban Sprawl": "Aesthetically, the result is a mess. It takes remarkably little blight to color a whole area; let the reader travel along a stretch of road he is fond of, and he will notice how a small portion of open land has given amenity to the area. But it takes only a few badly designed developments or billboards or hot-dog stands to ruin it, and though only a little bit of the land is used, the place will *look* filled up."[2]

In 1993, James Howard Kunstler voiced a similar outcry against the visual clutter of sprawl in his city, Saratoga Springs, New York:

> By any standards, South Broadway looks terrible. No thought has gone into the relationships between things—the buildings to each other, the buildings to the street, the pedestrian to the buildings. . . . The absence of trees planted along the sides of the street lends it a bleak, sun-blasted look, which the clutter of signs only aggravates. . . . The fast-food strip follows, all the little cartoon eateries in a row: McDonald's, Dunkin' Donuts, Long John Silver, Pizza Hut, Kentucky Fried Chicken. As a sort of crescendo to this long avenue of junk architecture, we arrive at the Holiday Inn . . . [with] all the formal charm of a junior high school.[3]

Clutter along streets and highways today has no more visual appeal than it did when Kunstler grumbled about it two decades ago (fig. 6.1). But not everyone hates sprawl. In his contrarian 2005 (pre-foreclosure crisis) book *Sprawl: A Compact History*, Robert Bruegmann describes critics (like Whyte and Kunstler, presumably) as driven by "a set of class-based aesthetic and metaphysical assumptions, almost always present but rarely discussed."[4] Bruegmann blandly dismisses race as a factor in driving sprawl, claiming that if "affluent enough to do so, African Americans have been just as willing as their white counterparts to move out to the suburbs."[5] He disregards, however, the pervasive redlining and "steering" of minority households away from white suburbs long after the 1968 Fair Housing Act banned such practices. With similar disdain he rejects contentions (like those made in chapter 3) that national policies encouraged sprawl: "The federal government, they say, fueled sprawl through homeowner subsidies, highway programs, infrastructure subsidies, and federal income tax deductions. Some anti-sprawl reformers go so far as to say that it was federal policies, not the private market, that all but forced tens of millions of Americans to live in the suburbs in single-family homes."[6]

While this assessment resonates with the tone of climate change skeptics, one may fairly agree that some early criticisms of sprawl based on aesthetics and tax costs to local jurisdictions seem quaint today. Yes, sprawl is ugly and wasteful of good farmland, forests, and scenic countryside. (Even those in the Bruegmann camp would be hard-pressed to quibble with the benefits of saving "sacred places"

Fig. 6.1. Visual clutter along arterial highway, Anywhere, USA.

like the Indiana Dunes.) But it is fair to admit that one person's ugly fast-food strip serves as someone else's source of a cheap meal and beverage to break up a tedious workday or highway journey.

Moving beyond what sprawl *looks like,* let's consider how metropolitan America *functions* for the four-fifths of Americans who now inhabit it. Present challenges can only worsen with the addition of another 100 million Americans by 2050, if estimates of the early 2000s are fulfilled.[7] Meanwhile, the nation's present population is aging and diversifying in race, ethnicity, and lifestyle: conventional two-parent families with children are a vanishing species; the increasing number of people of color will challenge conventional assumptions of the past era of the white, business-oriented establishment. Metropolitan America today confronts deteriorating infrastructure, shortages of affordable housing convenient to employment, crippling traffic congestion, an epidemic of respiratory disease and obesity, looming water shortages, increasing natural disaster costs, and global warming. While these problems, other than the last, vary from one region and nation to another, the overall prospect for the planet is, in Thomas L. Friedman's phrase, "hot, flat, and crowded."[8]

The Mismatch of Demography and Housing

In terms of cost, size, and location, the U.S. housing supply is perversely ill suited to meet the needs of a changing American population. Between 1940 and 2000, the nation slightly more than *doubled* in population (from 131 million to 281 million) while its total housing supply *tripled* (from 37.3 million to 115.9 million houses and apartments).[9] This would seem to indicate that the housing supply more than kept pace with population growth. But taking a closer look, it becomes apparent that housing supply and demography are woefully mismatched.

Suburban housing of the Levittown and Park Forest era (1940s and 1950s) provided needed starter homes for young (white) families with two parents and children. But as the century wore on, the traditional American household began to change. Average household size shrank from 3.68 in 1940 to 3.11 in 1970, and 2.59 in 2000.[10] By 2000, only 27 percent of all suburban households were married couples with children, outnumbered by "nonfamily" suburban households (29 percent).[11] In 2010, only one-fifth of all households were traditional families with children, and married couples had become a minority, accounting for only 48 percent of American households, as compared with 78 percent as recently as 1970.[12]

Another demographic change relevant to housing supply is the graying of America, with postwar baby boomers reaching retirement age and younger households having fewer or no children. The "over sixty-five" cohort grew from 8.1 percent of the population in 1950 to 13.0 percent in 2000 (reaching 40 million)[13]—a level that will continue to rise rapidly as the "boomers" reach the retirement threshold. While many older people choose to "age in place" in suburban ranch homes and split-levels acquired when their families were young, few are likely to buy into tract subdivisions of outer suburbia.

Meanwhile, offspring are marrying later, if at all, and having children, if any, later than their parents did, and thus represent a smaller market for traditional suburban housing. The rising proportion of Hispanic, Asian, and African American households has somewhat offset dwindling demand from white families, helping to broaden diversity in certain suburban markets.[14] But many young adults of all ethnicities were priced out of home ownership during the bubble years, at least in hot real estate market regions like New York and California. They became ready victims of the predatory loan industry, bringing on the foreclosure crisis since 2006. Abandoned foreclosed homes across the country today blight suburban neighborhoods that were yesterday's lower-cost dream homes on the urban fringe, as in Riverside and San Bernardino counties in California. And in remaining hot markets like San Francisco, a mecca for social media start-ups, longtime residents and businesses are being priced out by a "second tech bubble."[15] (As a supreme irony, the modest Greenwich Village row house once owned by Jane Jacobs—the high priestess of urban populism—sold for more than $3 million in 2010.)[16]

With the entry of women into the labor market and professional careers, the Ozzie and Harriet suburban stereotype mercifully wilted. By 2000, fully 70 percent of families with children were headed by two working parents or by an unmarried working parent.[17] Indeed, two incomes often were essential to cover the costs of housing that rose twice as fast as inflation between 1970 and the early 2000s, as well as the costs of two or more cars.[18] But Karen Kornblum, writing for the *Atlantic Monthly,* notes the power of social inertia: "The nation clings to the ideal of the 1950s family; many of our policies for and cultural attitudes toward families are relics of a time when Father worked and Mother was home to mind the children."[19] The perpetuation of the 1950s suburban ideal is nowhere more apparent than in the prevalence of single-family homes in subdivisions scattered all over the exurban fringe, far from sources of employment—testament to the resistance to change of local zoning codes and the building industry.

So what has the market provided to meet the needs of the increasingly older, childless, and unmarried American public? Answer: bigger, costlier, more garish, and more isolated versions of the same old single-family "dream house" on the metropolitan fringe. Earlier in the last century, the average home was 700 to 1,200 square feet on a 6,000-square-foot lot (about a seventh of an acre).[20] In 1950 the average home was 923 square feet,[21] which increased to an average size of 1,660 square feet in 1973 and 2,392 square feet in 2010,[22] during which year about 9 percent were larger than 3,200 square feet, some much larger.[23] Average lot sizes expanded during the postwar decades but with large variation from one region and housing market to another. Some of the smallest new subdivision lots (highest density) during the 1990s were located in California and Sun Belt cities, where land is extremely expensive due to constraints of physical geography and federal or tribal lands.[24] Average lot sizes nationally have declined slightly during the 2000s, possibly reflecting the influence of the Smart Growth movement and New Urbanism on local zoning regulations and development practices.

The bête noir of anti-sprawlists, of course, is the "McMansion," a product of the new Gilded Age of the 1980s through the mid-2000s. The McMansion is to the average home as the Hummer is to the VW Beetle. With several thousand square feet of living space, the McMansion exudes ostentation on a grand scale: cathedral ceilings, five or more bedrooms and bathrooms, a master suite with Jacuzzi, a kitchen equipped to prepare haute cuisine, a media room, a three-or-more-car garage, even an indoor swimming pool. Frequently, such oversized homes are situated in gated communities—subdivisions literally or psychologically insulated by fences, walls, gates, and even guardhouses against intrusion by unauthorized outsiders.[25]

McMansions—like the Gilded Age "cottages" at Newport, Rhode Island, minus their McKim, Mead & White elegance—may soon become costly relics of another age of greed and excess. Several factors could undermine their future appeal: the accumulation of great wealth in the hands of an ever-smaller percentage of the population; the cost of energy needed to heat and cool such a home (and fill up the Hummer as well); cultural revulsion against blatant bad taste; a decline in "flipping"—purchasing real estate as an investment, not for personal use, to be

resold at a profit—and a growing preference for more urban, higher-density surroundings. (Why live on a former cornfield if you can afford a condo in Brooklyn's
Park Slope or San Francisco's Marina District?)

Even without the Great Recession, the private market and its government
enablers have created a glut of oversized homes. The demographer Arthur C.
Nelson forecasts "a likely surplus of 22 million large-lot homes (houses built on
a sixth of an acre or more) by 2025—that's roughly 40 percent of the large-lot
homes in existence today."[26] The urbanist Christopher Leinberger, writing for
the *Atlantic Monthly*, asserts that even those who can afford a McMansion have
become "disillusioned with the sprawl and stupor that sometimes characterize
suburban life. . . . it is urban life, almost exclusively, that is culturally associated
with excitement, freedom, and diverse daily life."[27] Like smaller vacant homes,
foreclosed and empty McMansions present huge costs and diseconomies to their
surrounding neighborhoods and communities through lost taxes, lack of upkeep,
and vandalism—thus further depressing their resale market.

More broadly and bluntly, the mismatch between housing development and demography globally caused the post-2006 crash in housing prices and loan defaults
in North America, according to the Canadian urban writer Jeb Brugmann:

> The underlying problem was a global financial system disconnected from the changing
> economics of a continent [North America] that was rapidly urbanizing with little sense
> of urbanism. . . . [Lenders] made sweeping assumptions about the economic health of
> the urban regions upon which their borrowers depended, even as [sprawl] was chang-
> ing those economics fundamentally. . . . [New homes] were much larger, more energy-
> intensive, . . . and located farther away from employment areas, increasing their owners'
> mobility costs. . . .
>
> The result was a declining net economic equation, which no one was systematically
> calculating. Builders and home buyers invested in sprawling suburban tracts or high-rise
> condominium developments without factoring the full costs of operation, maintenance,
> and lifestyles, and how they would hit borrowers' pocketbooks along with high preda-
> tory mortgage rates.[28]

Meanwhile, even as millions of large and small homes fall vacant in the foreclosure debacle, there remains a stark, long-term crisis of housing affordability.
While the housing industry has built far too many monster houses for the upper
5 percent of the population that claimed 63.5 percent of the total wealth of the
United States in 2009, four-fifths of Americans collectively held only 12.8 percent
of national assets.[29] Obviously, many of the latter, especially the very poor and
working class households, are either priced out of metropolitan housing markets
or forced to pay a burdensome proportion of their income to own or rent a home
within reach of their places of employment.

This is not a new phenomenon, nor has it been alleviated much by the housing
downturn. The Center for Housing Policy reported rising levels of severe housing-

cost burdens to working households despite the recession. Of 46.2 million such households in 2009, nearly one-quarter pay more than half their incomes on housing, either ownership or rental, which translates into an increase of 600,000 (over a one-year period) in the number of households that were "severely burdened" in 2008.[30] Even in states with the highest foreclosure rates, the share of working households forced to pay over half their income for housing remained well above the national average of 22.8 percent, for example California (33 percent), Nevada (28 percent), Arizona (25 percent), and Florida (33 percent).[31]

The affordable housing crisis is both a result and a cause of suburban sprawl. On the one hand, the closer housing is to employment centers (downtown or outlying), the higher the cost to employees, many of whom may have to pay more than the recommended standard (30 percent of their income) to own or rent in such markets. On the other hand, to avoid the crippling cost of convenient housing or to house larger families, the trade-off is to live at the metropolitan periphery and incur long and costly commutes. An additional trade-off in either case has increasingly been to succumb to predatory lenders in the hope (now dashed) that house prices will rise to offset future interest rate hikes on adjustable mortgages.

Households of modest means are virtually priced out of high-cost housing markets today, as documented in a recent study of metropolitan Boston by the Urban Land Institute.[32] The study graphically relates household income status and housing costs in relation to selected employment centers in the Boston area. The resulting zones of "unaffordability" (i.e., costs of more than 30 percent of household income) occupy progressively larger swathes of the region as income level declines. For households at 60 percent of the area mean income (AMI), little if any affordable housing is available within an estimated forty-five-minute rush-hour drive from potential workplaces. With larger family size the picture is even bleaker, since closer-in housing is largely limited to one or two bedrooms. The study also sees no relief ahead as older housing deteriorates and new market-priced construction is generally out of reach to working families. Of course, households below 60 percent of AMI—the working poor and the unemployed—have few options apart from living with family, house-sharing, or dependence on dwindling federal rent vouchers, public housing, or homelessness.

Carmageddon: The Scourge of Traffic Congestion

In July 2011, Los Angeles shut down Interstate 405, for fifty-three hours over one weekend, to widen it. The highway connects western Los Angeles to the San Fernando Valley, with an average weekend traffic count of 500,000 vehicles.[33] The prospect of losing, albeit briefly, one of its key arteries, sent the city into high alert. Like Y2K (the belief that computers would shut down at the turn of the millennium), the closing itself, quickly dubbed "Carmageddon," was rather anticlimactic since people adjusted accordingly. But carmageddons occur daily in the nation's large metropolitan areas, exacting heavy costs in terms of time delays, wasted fuel, air pollution, and emotional distress.

Overcrowding of the nation's metropolitan highways, with its rising human, economic, and environmental costs, is in part a direct result of the mismatch between housing and demography described above. The journalist Neal Peirce expands the point: "The lack of reasonably priced housing close to employment centers is forcing more and more low- and middle-income families to choose locations far out in the countryside. 'Drive until you qualify,' goes the saying. But there's a heavy price to pay: vehicle miles traveled escalate, highway expenditures shoot up, air pollution increases. Duplicative schools and public services are required."[34]

The need to cover a lot of ground traveling from home to work and to all the necessities of daily life certainly contributes to traffic snarls, but congestion continues to be driven by the sheer proliferation of vehicles of all sizes and uses on the nation's deteriorating highways. In 2005, there were an estimated 237 million passenger cars, SUVs, light trucks, and motorcycles in use, as compared with 106 million in 1970, a 120 percent increase and nearly *three times the rate of total population growth* during that period. SUVs and light trucks alone increased near six-fold. The total of heavy trucks nearly doubled from 4.5 million in 1970 to 8.4 million in 2005.[35] Traffic congestion, of course, is also complicated by accidents, weather, road construction, and time of day. Even "rush hour" is increasingly meaningless in our largest metropolitan regions: I recently experienced mind-numbing, stop-and-go traffic conditions in late morning on I-10 westbound from Riverside to Pomona, California, fifty miles east of downtown Los Angeles for no evident reason other than too many people going from one place to another all day long.

Much of the interstate highway network is now more than fifty years old—as long as the period from the assassination of Abraham Lincoln until the outbreak of World War I, or from then until the Beatles era. Two generations of Americans have little or no recollection of life before the ubiquitous interstates. But Futurama's utopian vision of seamless, high-speed highways smoothly linking the nation's cities and countrysides is as outdated today as Delores Hayden's "sitcom suburb" (so is the sitcom itself a relic, at least as it existed in the 1950s). Ever-higher traffic loads and deteriorating road and bridge conditions threaten to paralyze mobility throughout entire metropolitan areas, and often between them as well. One *New Yorker* writer described a "bad day" in New York City in 2002 when two Hudson River bridges (George Washington and Tappan Zee),[36] and the primary route to New England (Interstate 95) all experienced accidents on the same Friday afternoon, producing regional gridlock.[37]

The infamous segment of Interstate 95 between New York and New Haven, one of the nation's oldest, and most overburdened highways, was originally planned for 70,000 vehicles per day. By 2002, its traffic count averaged 150,000 cars and trucks daily, in slow motion.[38] Such over-dependence on aging traffic arteries causes cascading havoc when a serious accident or structural failure occurs. For instance, in March 2004, a truck overturned and caught fire on I-95 in Bridge-port, Connecticut, causing an overpass to buckle and a "traffic nightmare" for the next couple of weeks.[39] A couple of hundred miles south, I-95 formerly crossed the Potomac River on the Woodrow Wilson Bridge, designed for 75,000 vehicles per day, and handling 190,000 as of 2000.[40] I-95 was rerouted around the city on the I-495 Capital Beltway, adding to the overloading of that notorious "parking lot." With congestion and deterioration come accidents: more than three thousand people died on I-95 during the 1990s.[41]

Growing traffic congestion, according to the Texas Transportation Institute,[42] inflicts huge costs on society. In 2009, "drivers wasted 5.7 billion gallons of fuel, or about 42 gallons per person" in the seventy-five largest urbanized areas. Traffic congestion now requires 3.5 billion hours of extra travel time and the economic costs of congestion are estimated by the institute at nearly $70 billion for 2009, an increase of $4.5 billion over the preceding year. The average annual cost per commuter was $808 in 2009 as compared with an inflation-adjusted total of $351 in 1982.[43] Time spent in traffic delays varies among urbanized regions, but is generally rising (see table 6.1).

6.1 Hours Wasted in Traffic Congestion (Average Annual Delay per Auto Commuter)

Urbanized Areas	1982	1999	2011
Atlanta	13	52	51
Chicago	18	55	51
Washington, DC	20	70	67
Dallas–Fort Worth	7	39	45
Boston	13	41	53
Houston	24	42	52
Seattle	10	52	48
Los Angeles	39	76	61
Baltimore	11	37	41
Average for 101 Largest Urban Areas	14	39	43

Source: Texas Transportation Institute, *2012 Urban Mobility Report*, excerpted from "Table 9. Congestion Trends 1982–2011," available at http://d2dtl5nnlpfr0r.cloudfront.net/tti .tamu.edu/documents/mobility-report-2012.pdf.

The prevailing approach to sprawl-induced traffic congestion has been to add more highway capacity through additional lanes or entirely new segments of highway. This approach has been counter-productive because increasing capacity attracts new demand, both from new development induced by the expansion of highways and by changes in motorist behavior in response to the perception of reduced congestion. This effect has been described as a "triple convergence" whereby new highway capacity causes drivers to abandon avoidance strategies of (1) avoiding peak times, (2) using other routes, or (3) using other modes of transportation.[44]

Modest local relief is gained from "tweaking" the highway system through such measures as bus and carpool lanes, ramp traffic signals, and accident management capabilities. In some metro areas such as Los Angeles and Washington, DC, some lanes are reserved for toll paying vehicles utilizing the transponders already used on conventional toll roads. The logical extension of toll-based highway usage is "congestion pricing" whereby differential fees are charged based on location, day and time, and other variables. This strategy avoids the triple convergence effect since drivers will not be lured back by a reduction of congestion if they have to pay for the privilege (and can choose to continue their avoidance behavior). Congestion pricing was inaugurated for central London in 2003 (based on its use in Singapore for twenty-five years) leading to immediate reduction in peak-hour congestion, as well as a new source of revenue potentially available to improve mass transit.[45] New York City mayor Michael Bloomberg shortly thereafter proposed congestion pricing for Manhattan, but suburban state legislators defeated the scheme. In 2006, the Partnership for New York City estimated that traffic congestion in and around the city cost the region over $13 billion a year in lost time, productivity, wasted fuel, and business revenue.[46]

Public Health: Asthma and Obesity

Ironically, suburban sprawl originated in the later nineteenth century with the flight of the wealthy from perceived threats to life and health that prevailed in cities—fires, crime, and infectious disease like cholera and yellow fever. The prevailing Victorian-era medical theory ascribed such outbreaks to "miasma," defined in this context as "the product of environmental factors such as contaminated water, foul air, and poor hygienic conditions.[47] It was believed that infection was not passed between individuals ("contagion") but would sicken people who resided in localities where such "vapors" (foul odors) were prevalent—to be avoided by escape to the nearby countryside. Cholera and many other infectious diseases were later ascribed to bacteria and viruses in contaminated water supplies[48] leading to the development of new hinterland water sources for Boston, New York, and Philadelphia, and other cities. But the prevailing sense that cities were inherently unsafe and unhealthy contributed to the exodus of the middle class to the suburbs beginning in the 1870s.

Now the shoe is on the other foot: it is the sprawled suburbs and the highways serving them that are public health hazards. Two areas of concern, among others, are respiratory conditions related to vehicle exhausts, especially diesel fumes from trucks and busses, and obesity resulting from the sedentary and stressful time spent driving from place to place, poor diet, and limited time or opportunity for exercise and active recreation.

As of 1999, about 15 million persons in the United States were estimated to suffer from asthma, resulting in more than 1.5 million emergency room visits, about 500,000 hospitalizations, and some 5,500 deaths annually.[49] The scourge of asthma has worsened since then: in 2011, the Centers for Disease Control reported that nearly one in ten children and one in twelve Americans of all ages now suffers from asthma. In 2009, 17 percent of black children had asthma, as compared with 11.4 percent in 2001.[50] The rise in reported cases is attributable to many factors such as allergens in the environment, quality and consistency of medical treatment, and genetics, but for the poor in particular, transportation-related pollution is perceived as a major cause. A joint study by the Natural Resources Defense Council and the UCSF School of Medicine in 2002 reported a strong connection between asthma and diesel fumes:

> Diesel exhaust from buses, trucks, other vehicles and equipment is a major source of sooty particles in the air. "There are a lot of sick people at the end of the tailpipe, especially children," said Dr. Gina Solomon, M.D., NRDC senior scientist and co-author of the study. The tiny particles in diesel exhaust bypass our respiratory defenses and lodge deep into the lungs. Once there, they stimulate an immune response that triggers inflammatory changes, airway constriction, mucus production and symptoms of asthma.
>
> Diesel exhaust also contains nitrogen oxides (NOx), a major component of smog. Although smog has been linked to asthma before, this study draws the strongest link to date between asthma and the small particles in diesel exhaust.[51]

Ozone is a major transportation-related cause of respiratory ailments in metropolitan areas like Atlanta, Los Angeles, and Chicago. Ground-level ozone forms when sunlight interacts with nitrogen oxides (from vehicles) and other atmospheric chemicals (also referred to as smog). In *Sprawl City*, a study of race, politics, and planning in Atlanta, Dennis Creech and Natalie Brown report that "high levels of ozone present one of the most serious air pollution problems. Negative impacts from ozone are most likely to be seen in children, athletes, the elderly, and people with preexisting lung disorders, but anyone exposed to ozone can be adversely affected."[52]

Asthma is not confined to inner-city highway corridors: it follows the traffic. As I mentioned in chapter 3, childhood asthma is endemic in the South Bronx, New York, where one in four children suffer from asthma today.[53] But in the outlying reaches of the greater Los Angeles region, children in Western Riverside and San Bernardino counties experience impaired lung function due to proximity of major highway and rail corridors serving the Port of Los Angeles, seventy-five miles away.[54] Over 700,000 California children are estimated to suffer from asthma related to highways and other diesel fume sources.[55]

Obesity, another scourge partly related to sprawl, is considered by many health experts and other researchers as among the most challenging of the health crises our country has ever faced.[56] "Obesity" is defined as a body mass index (BMI, a person's weight divided by the square of their height) exceeding 30; "overweight" is defined as a BMI exceeding 25.[57] The rise in these indicators is frightening: in 1990, no state had an obesity rate exceeding 15 percent of its population; by 2011, obesity rates exceed 25 percent in more than two-thirds of states. In 2000, only two states had a combined obesity and overweight rate above 60 percent; by 2011, forty-four states exceed that level. According to David Satcher, Surgeon-General of the United States: "Two-thirds of adults, 190 million people—are overweight or obese; nearly one third of children and teens fall into these categories."[58]

The health implications of being overweight are dire. Obese persons are four times as likely to incur Type 2 Diabetes (adult onset). Additional potential health impacts include heart disease, stroke, breast and colon cancer, and arthritis. Heaviness also imposes additional stress on joints, causing lower back pain and hip, knee, and ankle problems.[59] A report posted on the website of the Trust for America's Health includes troubling statistics: "Since 1995, diabetes rates have doubled in eight states. Then, only four states had diabetes rates above 6 percent. Now, 43 states have diabetes rates over 7 percent, and 32 have rates above 8 percent. Twenty years ago, 37 states had hypertension rates over 20 percent. Now, every state is over 20 percent, with nine over 30 percent."[60] According to the Harvard public health researcher Anne C. Lusk: "Sixty-five percent of the U.S. population is now overweight, and the resulting negative health consequences include premature death, cancer, heart disease, diabetes, stroke, and other chronic diseases. This rise in obesity is a result of poor diet and physical inactivity of an energy imbalance from an increase in caloric intake and a decrease in physical activity. Such activity provides a variety of physiological and psychological health benefits; . . . thirty to ninety minutes of moderate physical activity most days of the week [is recommended by the U.S. Department of Health and Human Services]."[61]

Children are particularly at risk from obesity. Richard Louv, in *Last Child in the Woods,* cites federal reports that by the year 2000, "two out of ten American children were clinically obese—four times the percentage of childhood obesity reported in the late 1960s.[62] (Surgeon-General Satcher, cited earlier in this section, reports that as of 2011 one-third of children and teens are obese and overweight.) But this rise in childhood obesity coincided with "a dramatic increase in children's organized sports."[63] Of course, closer analysis might distinguish low-income communities with high obesity rates (and low participation in sports) from places where the opposite is true. Nevertheless, Louv makes a convincing argument that—along with poor diet, and too much time in cars and in front of TVs or computers—lack of access to natural areas where children can play freely and energetically is a huge detriment to children's physical and mental health. (Organized sports may in fact provide little activity for those waiting in the dugout or on the sidelines.) He attributes this "nature deficit disorder" to several factors, including loss of natural greenspaces in suburbs, preference for pavement and artificial turf for outdoor recreation, fear of strangers in wooded areas, liability concerns of school and park managers, and lack of free time available to "containerized kids" in the lifestyle of suburban sprawl.[64]

The need for urban parks to better encourage vigorous exercise by adolescents and adults was addressed in a recent report from the Trust for Public Land Center for City Park Excellence.[65] Citing examples from cities as diverse as New York, Cincinnati, San Francisco, Denver, Minneapolis, and Seattle, among others, the report argues that "the mere presence of a park does not guarantee a healthier population. . . . With a growing clamor from doctors, parents, overweight persons, and even those who just want to strengthen muscles, lungs, and hearts, it's time for parks to be more than simply pretty places."[66]

Of course, the sedentary habits imposed by the sprawling metropolis threaten the health of its entire populace, not just kids. An American Cancer Society study tracked the activity patterns and health of 123,000 Americans between 1992 and 2006: men who spent six hours or more per day sitting had a death rate 20 percent higher than men who sat three hours or less. The death rate for women who sat more than six hours was 40 percent higher than their more active contemporaries.[67] According to the Mayo Clinic researcher James Levine, "Excessive sitting is a lethal activity."[68]

Traffic congestion, as I discussed above, adds up to dozens of hours per year standing still on many metropolitan freeways. But even when traffic moves smoothly, the time spent in vehicles and the number of miles driven continue to grow as well. Between 1960 and 1997, vehicle miles driven by passengers cars increased by 250 percent.[69] As of 1999, the Sierra Club estimated that the average American driver spent 443 hours each behind the wheel, the equivalent of fifty-five nine-hour days, or eleven workweeks.[70] And despite the proliferation of entertainment and communication gadgetry, driving in metropolitan traffic is mentally and physically stressful. ("Will that light ever turn green?")

No doubt then that driving, as a major side effect of sprawl, has a negative impact on public health, but the "design" of sprawl itself contributes to the impact as well. Metropolitan growth has unfolded as a patchwork of residential, retail, office, and industrial land uses, connected only by highways with little or no provision for pedestrians or cycling. How often does one have to drive from one mall to another across an intervening highway and acres of parking, or from a budget hotel to the nearest restaurant—visible but not easily or safely accessed on foot? The segregation of residential and business districts through zoning and development practices ensures that homes, shopping, health services, recreation, and other needs are isolated from one another.

Water Resources and Natural Disasters

Sprawl profoundly alters anything wet that falls within its path: saltwater estuaries and harbors; freshwater streams, lakes, ponds, and wetlands; and groundwater aquifers. The greatest percentage of stream miles affected by urbanization comprises dense networks of local streams, creeks, brooks, runs, kills, and bayous. Although modest in drainage area and average flow, local watersheds typically contain a complex array of hydrological and ecological elements, including headwaters, channels, banks, floodplains, lakes and ponds, wetlands, aquatic and riparian biotic habitats, aquifers, and coastal estuaries.[71]

Cities and local drainage networks are historically interdependent. Cities originated where a river or stream provided access to the interior from a river or harbor port at its mouth, as with New York City and the Hudson River valley or with New Orleans and the Mississippi. But much smaller streams fostered the growth of many cities, as with Hartford's Park River (now partly buried and forgotten), Boston's Charles River (now revived and revered), or Houston's Buffalo Bayou (now in process of recovery). The growth of port cities deepened harbors, hardened shorelines, and poured wastes into nearby waterways. Over time, the spread of pavement and buildings replaced natural land surfaces, curtailing natural seepage into the ground or nearby streams (see fig. 4.2). Storm sewers convey the unwanted runoff directly to local streams, polluting them and making flood events more intense and hazardous.

Between the 1930s and the 1970s, the nation's answer to floods was to pour concrete and more concrete; over 900 local flood control projects involving some 260 dams and reservoirs, over 6,000 miles of levees and floodwalls, and 8,000 miles of stream channelization were constructed by the Army Corps of Engineers and other agencies (fig. 6.2).[72] The lower Los Angeles River, famously, was encased in a concrete channel to protect or promote real estate development in its floodplain, a decision Mike Davis discussed in his 1998 book *The Ecology of Fear*: "Beneficial to large landowners, this strategy would force the natural river into a concrete straitjacket—destroying the riparian ecology and precluding use of the riverway as a greenbelt."[73] (I discuss efforts to restore portions of the Los Angeles River project in chapter 9.) Elsewhere, streams prone to flooding downtowns were entombed in tunnels as with Hartford's Park River and the Providence River in Providence, Rhode Island (which was "daylighted" and lined with a riverfront walkway and mini-parks in the early 1990s.)[74]

Fig. 6.2. Flooding of development behind flood control levee, Jackson, Mississippi, 1979.

In addition to flood control, water bodies in the path of sprawl have been re-engineered by navigation projects, hydropower development, highways, the dredging and filling of wetlands, off-stream diversions, waste discharges, and bank stabilization. About 40 percent of stream miles of the nation's waters assessed under the federal Water Act (including most urban waterways) are classified as "impaired"[75] in terms of their capacity to reduce flooding, filter pollutants, recycle nutrients, recharge groundwater aquifers, provide recreation and scenic enjoyment, and nurture aquatic biodiversity.[76]

Environmental justice issues abound in urban watersheds. Construction of low-income housing above or near local streams that were buried in the past to alleviate flooding have fostered neighborhoods of poverty, public health hazards, home abandonment, and shame, as with Philadelphia's Mill Creek "black bottom" studied for two decades by Anne Whiston Spirn and her students.[77] Externalities of upstream sprawl—including sediments, chemicals, debris, and increased flooding—impact older and less affluent communities and neighborhoods downstream. The Anacostia River, for instance, as the local historian John

Wennersten describes, flows from wealthy Montgomery and Prince George's counties in Maryland through distressed neighborhoods of Washington, DC, on its way to the Potomac:

> From the Second World War to the 1970s, the capital's problems with water increased exponentially. Many of these issues fell upon Anacostia and communities east of the river, which by the 1960s were undergoing demographic change. The watershed flooded constantly, causing damage and creating inconvenience for residents along the river and in the expanding suburbs of Prince George's County. The decision to use the Anacostia stream valley as the major automobile commuter corridor split the capital by bringing "white men's highways through black men's bedrooms." It also made the watershed an environment held hostage to a transportation grid.[78]

Wetlands are another widespread victim of sprawl. Many fresh and saltwater wetlands have been dredged and filled to create building sites; others have been polluted with runoff from sewers, industry, and landfills. Still others were intentionally flooded to create artificial water features to enhance surrounding development. Nationally, the loss of wetlands due to such activities was estimated in the early 1990s to be 290,000 acres per year. The bulk of this was for agriculture, but at least 13 percent (37,000 acres per year) was lost to suburban sprawl.[79] The loss of upstream wetlands aggravates downstream flooding due to loss of "natural storage" capacity, and degrades biotic habitats that support fisheries, bird populations, and other wildlife.

Sprawl also threatens the quantity and quality of groundwater, the primary drinking water source for one-third of Americans. During droughts, urban sprawl reduces percolation into aquifers by diverting run-off into sewers and thence to streams. Before federal and state clean water legislation in the 1970s, sprawl widely degraded ground and surface waters with industrial wastes, sewer outfalls, faulty septic systems, sediment from construction projects, highway salt, and other contaminants.[80]

In sensitive and hazardous settings like Southern California, upscale residential sprawl has widely invaded mountainsides, forests, deserts, chaparral, and coastlines in pursuit of scenic views, rusticity, and natural beauty. In privatizing spectacular views and "wilderness," the development process itself destroys them through access road construction, wildlife habitat disturbance, retail and resort expansion, and the proliferation of "showcase" homes cluttering up the expensive views of their counterparts. Although sprawl diminishes the pristine beauty, biodiversity, and isolation that wealthy high-rollers crave, it also triggers a series of interconnected hazards and disasters that Mike Davis has observed: "The extreme events that shape the Southern California environment tend to be organized in surprising and powerfully coupled causal chains. Drought, for example, dries fuel for wildfires which, in turn, remove ground cover and make soils impermeable to rain. . . . In such conditions, storms are more likely to produce sheet flooding, landslides, and debris flows that result in dramatic erosion and landform change."[81]

Disregard of natural hazards by developers and local governments has been unintentionally encouraged and subsidized by federal disaster assistance.[82] Even when repetitive disasters strike the same area, rebuilding more lavishly, with the aid of private insurance and public disaster assistance, is the usual outcome. Thus the East Bay Hills overlooking San Francisco Bay—and overlying the hazardous Hayward Fault—were rebuilt with much larger (albeit more fire-resistant) homes after three thousand structures were destroyed in an October 1991 wildfire. Despite proposals to acquire the burned area for public open space, local governments like the City of Oakland vowed to rebuild the Hills (as their prime tax base) with federal disaster assistance enabling the process—thus setting the stage for even more costly conflagrations in the future.[83] Similarly, sprawling development along coastal barrier islands and other fragile shorelines is typically replaced with federal assistance after a disaster strikes, despite the ongoing risk of erosion and flooding.[84]

Fragmented Governance

Sprawl is both a cause and effect of political fragmentation. As I noted in chapter 4, the governance of metropolitan areas is sliced and diced among a vast and random crazy-quilt of municipal governments and counties, each endowed by state law with authority to levy property taxes, to plan and zone land use, to promote commerce, to establish and maintain public facilities such as police and fire stations, schools, parks, and so forth. Municipal governments come in all shapes, sizes, and political cultures. Yet from central cities like Chicago to pseudo-rural suburbs with names like "Country Club Hills," each is established as a separate legal and political fiefdom within the overall balkanized metropolitan region (fig. 6.3).

Legally, American cities and other units of local governance are all remote descendents of the medieval municipal corporation—a city awarded certain powers of self-government under a royal charter. Then and now, municipal powers and economic growth went hand in hand. In medieval times, merchants and artisans through their trade guilds, like modern trade associations or chambers of commerce, were closely linked with local political authority. In postwar America, city government and corporate powers formed a unity of purpose that harnessed large-scale public infrastructure projects (e.g., urban renewal, highways, convention centers, sports stadiums, airports) to the attraction and retention of corporate investment, thus enhancing employment, tax revenue, and civic prestige.

There are about twenty-five thousand general purpose municipalities in the United States today, better known as cities and towns, along with—depending

Municipalities: Seven-County Metropolitan Chicago Area

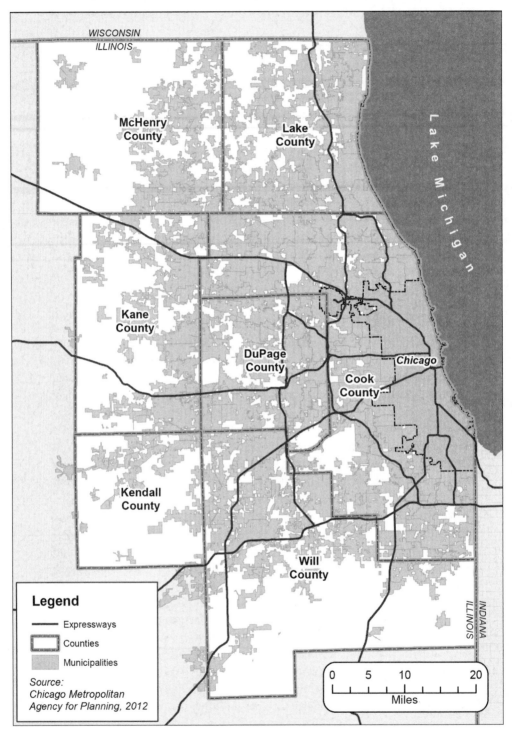

Fig. 6.3. Metropolitan Chicago is fragmented among hundreds of local zoning and taxing units of government. (Courtesy of the Chicago Metropolitan Agency for Planning).

on the state—counties, parishes, boroughs, townships, and incorporated villages. In many metro areas, regional public services such as schools, water and sewer facilities, mass transit, and airports are operated by special districts and authorities established to spread the costs and benefits of services across a wider regional public. Certain metropolitan counties like Montgomery County outside Washington, DC, Cook County in Illinois, and Los Angeles County operate at their own government level to provide various services to all or parts of their areas of jurisdiction.

Further complicating the geography of the post-sprawl metropolis are layers of special districts and authorities that provide specific services such as mass transportation, water and sewage, solid waste collection, parks, and other urban needs. These entities—nowhere marked on your AAA road map—often respond to (often in an effort to compensate for) the dysfunction inherent to "general purpose" municipal government. As quasi-governments, special districts build and operate public infrastructure with revenue from bond issues, taxes, and fees for services. Ideally, they provide an efficient means to allocate the costs of public services to those who benefit from them, and also to remove the administration of vital public services from the reach of local politics and patronage. That goal, however, may be elusive as special districts themselves become politicized over time.

Most special purpose governments are obscure to the general public and thus may lack political accountability. Some of them are exemplary, such as the Massachusetts Water Resources Authority, created in 1985 to update Boston's metropolitan area water and sewer systems, and to clean up Boston Harbor—goals that have largely been achieved.[85] But many special units of government have been created around the country to evade the debt limits on conventional governments and to construct financially risky mega-projects like stadiums, convention centers, and redevelopment ventures that are shunned by private enterprise. When faced with default, heavily indebted local and state governments may be forced to satisfy the obligations of the special authorities they have created. Such was the case with the New Jersey Sports and Exposition Authority which failed to repay bonds that financed the Giants Stadium and a racetrack in the Meadowlands across the Hudson River from New York City, forcing the state to assume the authority's interest payments of $100 million a year.[86]

Federal state, and local authorities thus fragment the governance of metropolitan America vertically, whereas myriad general purpose and special purpose units split it horizontally. This situation has resulted from permissive state laws regarding incorporation of both general and special-purpose governments as well as from a corresponding failure to provide for consolidation or termination of any of them. In an era of shrinking public revenues and resistance to tax increases, the shaky finances of many local municipalities and their special district surrogates is the governmental outgrowth of the foreclosure crisis in the real estate market.

Regionalism has been a Holy Grail for planners and civic leaders since the 1902 Washington, DC, McMillan Commission Plan, the 1909 *Plan of Chicago,* and the *Regional Survey of New York and Its Environs* of the late 1920s. The work continues today through flagship regional agencies like New York's Regional Plan Association, Boston's Metropolitan Area Planning Commission, Chicago's Metropolitan Agency for Planning, the Denver Regional Council of Governments, the Association of Bay Area Governments, and their counterparts in other cities. But decentralization and localism, as noted by Jon C. Teaford, have always been "a sacred element of the American civil religion and the nation's lawmakers [are] devout in their adherence to the faith."[87] That shows little sign of changing in the foreseeable future.

Climate Change

In the summer of 2011 (as I am writing the first draft of this chapter), an extraordinary heat wave has blasted states from Arizona to Florida and from Missouri to Louisiana with weeks and months of triple-digit temperatures. On the very day I write (August 4, 2011), about twenty-five southern cities are expecting their highest temperatures ever recorded, for example: Oklahoma City (104 degrees), Wichita Falls, Texas (109), Little Rock, Arkansas (108), Yuma, Arizona (116).[88] (The day before, Texas set an all-time record for electrical power demand, prompting calls from state energy officials for voluntary cutbacks to avoid brownouts.) The costs of this heat wave and drought cannot yet be assessed, but they will include immense impacts on agriculture, urban and rural water supplies, energy demand, air quality, ecological habitats and wildlife, outdoor recreation, and human health and well-being in the affected regions. (Meanwhile, a far more devastating drought in the Horn of Africa is inflicting famine, dehydration, infectious disease, and infant mortality as I write.)

The 2011 heat waves provide a foretaste of the potential impacts of climate change worldwide. Atmospheric concentrations of carbon dioxide—a signature greenhouse gas along with methane—have increased from about 280 parts per million (ppm) at the start of the Industrial Revolution to 370 ppm by 2000, with the increase from 1960 to 2000 of 54 ppm far greater than the 36 ppm rise from 1760 to 1960.[89] As of May 2013, carbon dioxide is flirting with 400 ppm at the Mauna Loa observatory in Hawaii, far from urban centers.[90] Gradual warming of the atmosphere and oceans is expected to intensify extreme weather events, including heat waves, droughts, tropical storms and flooding, and even winter storms. Furthermore, the well-documented melting of mountain glaciers in the Andes, the Himalayas, the Alps, and the Rockies—Glacier National Park will be almost ice-free in a few years—is potentially devastating to the billions of people worldwide who depend on glacier-fed rivers for navigation, hydropower irrigation, water supply, and fisheries. Sea levels are gradually rising due in part to the melting of both alpine and polar glaciers. The *New York Times* reported November 13, 2010: "Within the past decade, the flow rate of many of Greenland's biggest glaciers has doubled or tripled."[91] Sea level rise is already attacking low-lying shorelines along river deltas, estuaries, and barrier islands, magnifying the damage from hurricanes and storm surge.

The rise in atmospheric carbon dioxide levels, and hence climate change, results from two primary sources, the burning of fossil fuels and deforestation. Lester R. Brown documented the following statistics for both sources in his 2001 book *Eco-Economy:* "Each year more than 6 billion tons of carbon are released into the atmosphere as fossil fuels are burned. Estimates of the net release of carbon from deforestation vary widely, but they center on 1.5 billion tons per year."[92] The largest sources of fossil fuel CO_2 are electrical power plants and motor vehicles. Brown also noted in 2001 that the "global fleet of 532 million gasoline-burning automobiles [and trucks], combined with thousands of coal-fired power plants, are literally the engines driving climate change."[93]

As this book goes to press (August 2013), the impacts of global warming—much of it due to rising emissions from motor vehicles around the world—are becoming dire. Chronic drought in the American Southwest has caused the wildfire season to lengthen by two months over the last thirty years.[94] With the rapid spread of exurban development into forest and mountain landscapes, combined with ongoing drought and die-off of millions of acres of pine forests from insects, wildfire is now an existential threat. In 1993, federal agencies spent about $140 million fighting forest fires on nearly 1.8 million acres of land. In 2012, the federal government spent ten times as much, about $1.9 billion, to fight fires covering 9.3 million acres.[95] An estimated 740,000 homes with a total value of $136 billion are located in the "Wildland-Urban Interface," thus confronted with extreme fire danger.[96]

Another climate-related catastrophe, Hurricane Sandy, struck the mid-Atlantic region of the United States on Halloween 2012, causing coastal and riverine flooding from North Carolina to Maine and as far inland as Ohio. Millions lost power for periods lasting up to several weeks. Storm surges up to thirteen feet on top of high tides engulfed coastal communities of New Jersey, New York, and Connecticut. In metropolitan New York, subway tunnels, commuter railways, electrical systems, and other infrastructure were widely disabled by flooding. Flooding of a power station on the East River blacked out the financial district and most of Lower Manhattan for several days. Bellevue Hospital lost its backup power, and patients had to be transferred to other facilities by columns of ambulances. Beachfront communities and parks experienced catastrophic flooding which reached several blocks inland, causing heavy damage to public boardwalks, recreational facilities, and adjoining residential and commercial structures. The disabling of several sewage treatment plants released raw sewage into floodwaters, adding to public health hazards throughout the region.[97] Residents of public housing projects near waterfronts were deprived of power, heat, elevators, and daily necessities for several weeks. The public and private costs inflicted by Sandy are expected to exceed $50 billion, surpassed (so far) only by Hurricane Katrina in 2005.

These are fearful portents of a "new normal" in disasters, both sudden and gradual, relating to climate change. The Nobel Prize–winning International Panel on Climate Change "has found with near certainty that human activity is the cause of most of the temperature increases of recent decades, and warns that sea levels could conceivably rise by more than three feet by the end of the century if emissions continue at a runaway pace."[98]

PART III

The (More) Humane Decades, 1990–Present

"How Ordinary Citizens Are Restoring Our Great American Cities"—the apt subtitle of Harry Wiland and Dale Bell's book (and popular PBS series) *Edens Lost & Found*—eloquently attests that "humane urbanism" is thriving at many scales in such large cities as Chicago, Philadelphia, Los Angeles, and Seattle. Proactive city mayors like Richard M. Daley (Chicago), Michael R. Bloomberg (New York), Anthony A. Williams (Washington, DC), and Antonio Villaraigosa (Los Angeles) have redefined the role of municipal governance. Where local governments once marched in lockstep with the drumbeat of federal, state, and corporate priorities, they now also respond to goals defined by community groups and NGOs on behalf of "ordinary communities," not just "downtown." For instance, *PlaNYC: A Greener, Greater New York*, the sustainability blueprint of Mayor Bloomberg, lists dozens of neighborhood-scale actions relating to affordable housing, parks, tree-planting, public transportation, and brownfield clean-up.

In *Small, Gritty, and Green,*[1] Catherine Tumber discusses how small cities like Holyoke (Massachusetts), Muncie (Indiana), or Toledo (Ohio)—as well as formerly large industrial cities like Detroit and Cleveland—are seeking to redefine themselves as centers of innovation and "low carbon" sustainability. Some are experimenting with new strategies for "smart decline," a plan to re-concentrate homes, businesses—and the public services catering to them—in viable neighborhoods while converting abandoned lots to community gardens and related activities.[2] There is no panacea: many such cities today face hobbling unemployment, housing foreclosures and abandonment, drug and alcohol abuse, and other economic and social afflictions. But programs like Holyoke's Nuestras Raices (community farming and business incubation), New Haven's Urban Resources Initiative (regreening of vacant lots), and the Marvin Gaye Greenway in Washington, DC (each summarized in chapter 10) offer inspiring glimpses of homegrown inventiveness in small cities.

The four chapters of "The (More) Humane Decades," the third and final part of this book, explore the terrain of contemporary humane urbanism from several perspectives. Beginning in the 1990s, the urban design professions sought to respond to the last century's sorry legacies through the Smart Growth movement and its cousin New Urbanism. Those initiatives have helped to broaden the geographic and functional scope of urban improvement beyond the preoccupation with downtowns, automobiles, and the single-use zoning of the past century. In particular, they have promoted more flexible regulations to encourage mixed-use development providing a range of housing types and costs, proximity of residential and commercial activities, and reduced dependence on cars through public transit, bikeways, and sidewalks.

Smart Growth and New Urbanism have thus helped to improve the quality of particular urban projects, such as the reuse of Stapleton Airport in Denver or the planned Atlanta BeltLine (I discuss both in chapter 7). But they have little or no traction across the vast swathes of metropolitan America that are already built and not targets of major investment anytime soon—what Joel Kotkin has termed "midopolis."[3] Even those areas of unassuming older neighborhoods, highway strip development, and decaying public facilities have begun to benefit from an array of new legal and financial tools—largely created or reinvigorated during the 1990s—which facilitate humane urbanism at various scales. Some of these "acronymic" devices summarized in the balance of chapter 7, "Replanting Urbanism in the 1990s: A Garden of Acronyms," include affordable housing strategies, the American with Disabilities Act, green buildings, rail trails, urban water resource management, endangered species, and climate change adaptation.

The final three chapters explore how these and other new strategies are helping to promote humane urbanism in varied metropolitan settings across the country. Chapter 8, "New Age 'Central Parks': Two Grand Slams and a Single," heralds the synergy of people, place, and nature achieved in Chicago's Millennium Park and New York's High Line, in contrast with Boston's disappointing Rose Fitzgerald Kennedy Greenway—a throwback to the era when parks were promised as bait to gain public support for urban highways (a strategy that failed with New York's Westway).

Chapter 9, "Reclaiming Urban Waterways: One River at a Time" compares strategies for rehabilitating urban streams and watersheds in the metropolitan areas of Boston, Washington, DC, Houston, and Los Angeles. Despite differences in physical geography, land use, politics, and demography, such initiatives depend on public-private collaboration as catalyzed by nongovernmental watershed organizations and local movers and shakers.

Finally, chapter 10, "Humane Urbanism at Ground Level," samples a few of the myriad efforts now in progress to reconnect people, place, and nature through such earthy and multi-objective pursuits as urban farming, urban tree planting, vacant lot reuse, greenway creation, wildlife refuge advocacy, and environmental education.

Chapter 7

Replanting Urbanism in the 1990s: A Garden of Acronyms

> *The city, suburbs, and the countryside must be viewed as a single, evolving system within nature, as must every individual park and building within that larger whole. . . . Nature in the city must be cultivated, like a garden, rather than ignored or subdued.*
> —ANNE WHISTON SPIRN, *The Granite Garden*, 1984

> *Three areas are each on the cusp of change: regionalism is a reality about to be born, the suburbs are rapidly maturing, and many inner-city neighborhoods are primed for rebirth. . . . The challenge is to clarify the connections and shape both the neighborhood and region into healthy, sustainable forms—into Regional Cities.*
> —PETER CALTHORPE and WILLIAM FULTON, *The Regional City*, 2001

The decade of the 1990s saw the beginning of the end for top-down urbanism described in parts I and II. The Patrician Decades with their well-meaning aesthetic pretensions had long receded into planning history textbooks by 1990, though many individual patricians continued to play key roles in devising and funding new urban agendas, such as the many urban greening projects spearheaded by New York's mayor Michael R. Bloomberg. Likewise, the postwar hegemony of technocrats that decimated the nation's older cities while driving the suburban sprawl engine receded in its unquestioned authority—most dramatically signaled by the demise of New York's Westway project in the mid-1980s, which I discuss in chapter 8.

No single earth-shaking event, however, marked the gradual "sea change" from top-down, command urban policies to a more grassroots, pluralistic, and humane urbanism. Rather, this subtle transition has involved a series of loosely connected new laws, policies, and paradigms ("garden of acronyms") beginning in the early 1990s whose effects have been unfolding to the present time. The Smart Growth and New Urbanist movements that evolved in the 1990s rode the crest of the breaking wave of frustration with the technocratic focus on the automobile, suburban sprawl, and separation of residential and commercial uses. While retaining an "expert-driven" perspective (as a marketing tool), Smart Growth arguably provided a segue from the "one-size-fits-all" era before the 1990s to the more homegrown approaches that have proliferated ever since.

Smart Growth and New Urbanism

With striking symmetry, the twentieth century opened and closed with a surge of urban evangelism. As the City Beautiful movement stirred mainstream progressivism in the early 1900s, Smart Growth and its cousin New Urbanism dominated the hearts and minds of urban practitioners at the century's close.

But Smart Growth, broadly speaking, is as different from City Beautiful as the computer age from the era of the telegraph. If experience is a demanding teacher, the "movers and shapers" of the nation's cities have absorbed many lessons since the heyday of technocratic urban renewal and suburban sprawl. The Smart Growth canon reflects the early insights of heretical "seers of the sixties" like Lewis Mumford, Kevin Lynch, William H. Whyte, Jane Jacobs, and Ian McHarg concerning sprawl, transportation, neighborhoods, open space, walkability, and scale. To these were added concepts from the 1970s and 1980s, such as growth management, affordable housing, mixed-use development, infill and brownfield redevelopment, transit-oriented development (TOD), low impact development (LID), and other strategies.

In a radical departure from the Technocrat Decades, Smart Growth America (SGA)—the movement's central dynamo based in Washington, DC—focuses on neighborhoods rather than "downtown":

> At the heart of the American dream is the simple hope that each of us can choose to live in a neighborhood that is beautiful, safe, affordable and easy to get around. Smart growth does just that. Smart growth creates healthy communities with strong local businesses. Smart growth creates neighborhoods with schools and shops nearby and low-cost ways to get around for all our citizens. Smart growth creates jobs that pay well and reinforces the foundations of our economy. Americans want to make their neighborhoods great, and smart growth strategies help make that dream a reality.[1]

There is a slight whiff of *Mister Rogers' Neighborhood* in this statement, but the goals that it reflects are a welcome contrast to the downtown fixation of city boosters of the past century.

In another contrast to earlier attempts to "manage growth," Smart Growth de-emphasizes land use regulation as a tool for achieving its mission. Invigorated by a conservative majority on the U.S. Supreme Court,[2] property rights advocates since the 1990s have stigmatized public regulation of land use as "command and control" (appropriating a Cold War term). In a quiet burial of the zoning gamers' efforts to challenge local exclusionary zoning (discussed in chapter 4), planners and public officials in the 1990s began to soft-pedal more restrictive zoning proposals, except to allow more flexibility in the design of new development.[3]

But a silver lining to the diminished role of land use regulation has been a greater stress on negotiation and mediation involving all stakeholders in preference to "top-down" solutions. Smart Growth actively seeks to enlist the development community and local government—the bêtes noires of past anti-sprawl crusades—as allies rather than opponents. Henceforth, the emphasis would be, not to slow or stop growth, but to guide it toward better locational and design results through partnerships of environmentalists, builders, local officials, and design professionals. SGA co-founder Donald Chen has written about this shifting focus in *Scientific American*: "As communities becomes dissatisfied with haphazard growth, they are rebelling against the conventional wisdom that continued sprawl is desirable, immutable and inevitable. Urban, suburban and rural residents have joined forces in coalitions that would once have seemed improbable."[4]

Many organizations, states, and local governments have adopted smart growth policies that emphasize their particular goals. For instance, the National Association of Home Builders (NAHB) "Statement of Policy on Smart Growth" identifies the following five goals: (1) meeting the nation's housing needs; (2) providing a wide range of housing choices; (3) a comprehensive process for planning growth; (4) planning and funding infrastructure improvements; (5) using land more efficiently; and (6) revitalizing older suburban and inner-city markets.[5] Strategies for achieving those goals include:

• Open space conservation.
• Urban growth boundaries.
• Compact, mixed-use developments.
• Revitalization of older downtowns and inner-ring suburbs.
• Viable public transit.
• Regional planning coordination.
• Equitable sharing of fiscal resources across metropolitan regions.[6]

The Congress for the New Urbanism (CNU), founded in 1993, has proselytized the development industry with its version of Smart Growth, an urban design paradigm featuring traditional walkable communities modeled on the commuter rail suburbs of the 1920s. New Urbanists advocate revising local zoning codes to allow higher densities, diversity of building styles, mixed-use neighborhoods, front porches, sidewalks, and (one assumes) protection of mature trees and patches of habitat.

New Urbanism, like the City Beautiful movement in its day, is first and foremost about architectural and community design, a priority expressed by Peter Calthorpe and William Fulton: "Put simply, the New Urbanism sees physical design—regional design, urban design, architecture, landscape design, and environmental design—as critical to the future of our communities. While recognizing that economic, social, and political issues are critical, the movement advocates attention to design. The belief is that design can play a critical role in resolving problems that governmental programs and money alone cannot."[7]

And according to CNU's co-founder and troubadour-in-chief Andrés Duany: "Whether it is street width, housing density, building placement or landscape layout, no design decision should come in isolation. This is the fundamental insight of the New Urbanists: paying careful attention to how the urban design coheres, drawing on the lessons of prewar developers."[8]

Celebration and Stapleton

Parallels with Ebenezer Howard's garden city movement and its American counterparts come to mind. But while garden cities were to be built through limited-dividend investment by public-spirited progressives to help England's working classes (or by the New Deal Resettlement Administration in the United States), New Urbanist model towns like Seaside and Celebration in Florida are upscale and modish alternatives to conventional subdivisions, created by for-profit companies. They represent a clever and, in some respects, a desirable marketing vision. But *New Urbanist* does not necessarily equal *urban*. The hurly-burly of real neighborhoods and downtowns revered by William H. Whyte and Jane Jacobs will not be found in New Urbanist communities like the Disney town of Celebration, Florida, where: "The public spaces, just like the commercial buildings, [are] owned and controlled by Disney and a long list of restrictions is written into the sales contracts for every house."[9]

In *How Cities Work,* Alex Marshall observed that older communities, like Kissimmee, Florida, down the highway from Celebration, reflected the economic and transportation context in which they evolved. An attempt to design a town in the automobile age to look like a pedestrian and rail-based community of the 1920s is inherently flawed: "What Celebration very much is *not,* is a cure to the sprawl around Orlando. If Disney wanted to help combat sprawl, the worst thing to have done was to build another subdivision twenty miles [from downtown Orlando]. . . . Celebration is an automobile suburb. It can do little to escape from that dynamic."[10]

In 2010, a *New York Times* op-ed column expressed a quasi-epitaph for Celebration:

> Celebration's initial design of a downtown core to emphasize walking over cars and friendliness over isolation started to disappear even before Disney ceded control. Ever-larger houses have spilled across hundreds of acres of reclaimed swamp, replacing the small-town feel with something closer to traditional suburban sprawl. . . . The foreclosures that has [*sic*] swept across Florida hit Celebration hard. One real estate web site recently listed 492 foreclosures in town, and housing prices have dropped sharply. . . . The movie theater, once a focal point of downtown, shut its doors on Thanksgiving Day.[11]

Celebration and the equally famous Seaside on the Florida panhandle are easy targets of skeptics.[12] But the hundreds of developments influenced by New Urbanism over the past twenty years are not readily dismissed. One of CNU's pronounced success stories has been its influence on the 1990s generation of public housing known as HOPE VI. The results can be startling: the front porches in a row of pseudo "farmhouses" in Bridgeport, Connecticut, forlornly survey a vast nothingness, but the houses are certainly an improvement over the failed high-rise projects of the past. In Holyoke, Massachusetts, a HOPE VI development called Churchill Homes (fig. 7.1) is the pride and joy of that city's public housing authority, and a pleasant contrast to the city's remaining pre-war housing project, Lyman Terrace (compare fig. 3.1).[13]

One of the largest and most promising Smart Growth / New Urbanist projects is the redevelopment of Denver's former Stapleton Airport. In 1995, Federico Pena, former Denver mayor and then U.S. transportation secretary, promoted the construction of Denver International Airport far out on the plains. While the new airport site was a flagrant case of government-induced sprawl, it released

Fig. 7.1. Churchill Homes, HOPE-VI low-income public housing project, Holyoke, Massachusetts.

the 7.5 square-mile Stapleton site near downtown Denver for reuse. Calthorpe Associates was retained by the developer Forest City Enterprises to design what would be the nation's largest New Urbanist project. The Calthorpe plan envisioned a $5 billion investment over fifteen years to construct a new community of some twelve thousand energy-efficient homes—eight thousand single-family dwellings and four thousand apartments—along with ten million square feet of office space, several shopping complexes, six public schools, and more than 1,100 acres of pub-lic parks and open spaces.[14] The airport's former runways would be recycled as material for streets and other paved areas in the new project—the nation's largest recycling project of its kind.

At this writing, the redevelopment of Stapleton is well underway despite the re-cession, with over three thousand homes completed of various prices and designs, more than eight thousand residents, two major retail centers, offices, schools, and hundreds of acres of parks, with much more to come.[15] A light rail station is planned to connect the new community to downtown Denver and other parts of the city. In 2004, the $14 million new Denver School of Science and Technology opened on the site. The Stapleton Foundation for Sustainable Development was established in 1995 to promote the economic and social goals of the project and build a sense of community.

Village Hill and South Dunn Street

When the Massachusetts legislature closed a historic state mental hospital in Northampton, Massachusetts, in 1992, it specified that the site be redeveloped for residential and commercial purposes, with 15 percent of the housing units to be earmarked for mental health clients. Calthorpe Associates, under contract with the state, proposed a New Urbanist plan involving a mix of housing types with retail and office uses, including reuse of certain historic buildings, to coexist with "a new mental health education center and a hotel with conference and banquet facilities. A traditional Main Street connects these elements and the surrounding town."[16]

Fast forward: The development of "Village Hill" (as the old hospital site was euphemistically renamed) as of 2012 is indeed providing a mix of housing types— about sixty affordable apartments in two renovated staff buildings, another forty or so new rental units, a growing number of townhomes and bungalow-style units, and a row of upscale single-family homes. This blend of energy-efficient residen-tial units—some rented and others owned, some subsidized and others at market cost—reflects the spirit of New Urbanism. However, Village Hill to date lacks any retail or community facilities, let alone a hotel or "mental health education cen-ter." A sizeable tract of land designated for commercial use was redeveloped by a defense contractor whose well-secured plant obstructs the formerly spectacular view of a nearby mountain range from the residential side of the development.

Fig. 7.2. Mixed-income new housing units at Village Hill, Northampton, Massachusetts.

An assisted-living corporation is considering locating in the development. But a "traditional Main Street" is nowhere on the horizon—Village Hill has numerous access roads bordered by sidewalks which few people use. The development does have occasional bus service and in good weather the more active residents may walk or bike into downtown Northampton a mile away (with a long hill climb on the way back). But Village Hill according to its website[17] is supposed to be a "vibrant community" whose mantra is "Community, Commerce, Culture"—all of which are so far conspicuous by their absence (fig. 7.2).

For purposes of comparison, I recently visited an inviting mixed-use New Urbanist development now in progress at South Dunn Street in Bloomington, Indiana. The several dozen homes completed and occupied so far line narrow streets bordered with tree belts and sidewalks. As per New Urbanist principles, single-car garages open onto alleys behind the houses with parking largely prohibited on the streets. Each house is painted with a different color scheme and is fronted with, of course, a traditional front porch. The overall effect is indeed charming and compatible with other nearby Bloomington neighborhoods. A row of two-story brick structures with storefronts below and apartments upstairs lies around the corner, housing a consignment shop and a beauty salon among other tenants.[18]

Unlike Village Hill, South Dunn Street is embedded into the fabric of the exist-ing town, with shopping, professional offices, and Indiana University all within easy walking or cycling distance (in good weather at least). Another contrast with Village Hill is the absence of "social engineering"—no units are subsidized or ear-marked for persons with mental or physical disabilities. This in fact is an upscale real estate development in a comparatively prosperous university town of seventy thousand people, and a favorable contrast to the more plebian tract developments further out of town. It is designed, located, and marketed to be a "safe investment," and it will presumably fulfill that ambition.

Comparing South Dunn Street with Village Hill highlights a contradiction in the New Urbanist ideology: private profit and public interest may be incompat-ible. South Dunn Street does not aspire to serve the under-housed; it simply offers shining new homes to those who can afford them in a prime location where any decently planned residential development would succeed (New Urbanist or not). Even a dearth of commercial tenants would not be detrimental to the homebuyers, since the rest of Bloomington is close at hand.

Village Hill by contrast was legislatively intended not only as a source of profit for developers but also as a socially responsible, mixed-use, and "vibrant" com-munity to be built in toto on a site that is a mile from downtown Northampton. But its primary amenity—a gorgeous view—was squandered by city planners who allowed a poorly located industrial building to block it, and the site is further bur-dened with the continued use of the former hospital administrative building for state offices with all the neighborly ambience of San Quentin Prison. While the hospital's principal building has been razed (over the objections of preservation-ists), other vacant former hospital buildings still dominate the site, too "historic" to be razed but awkward to adapt for new uses. So new homes are sprouting amid remnants of the old hospital, and their residents, except the very fit, must still drive to anywhere else in town.[19]

Smart Growth and New Urbanism represent a *necessary* but not *sufficient* response to suburban sprawl and central city decline. New Urbanism in particular has little traction for the vast swathes of older cities and suburbs not targeted for major infill investment, areas dubbed "midopolises" by Joel Kotkin "that have become less a frontier of development than a shifting middle ground between the urban core and new growth nodes along the metropolitan edge."[20] The "midopolis" by definition has been largely untouched by major new growth or infill devel-opment. Vast swathes of Rust Belt cities like Detroit, Philadelphia, Buffalo, Bal-timore, and St. Louis have notoriously lost population, jobs, housing, and retail businesses. Nationwide, the subprime mortgage collapse has destroyed homeown-ership equity and ravaged older and newer communities alike: foreclosure epicen-ters in the Sun Belt and older industrial cities are checkerboarded with abandoned and decaying properties, dragging down values of still-occupied homes nearby, and causing huge property tax shortfalls to local jurisdictions. These are tough places and times to preach New Urbanist dogma: front porches and walkability are nonstarters if the neighborhood is dead or dying.

But let us not judge the importance of Smart Growth strictly by the design prescriptions of its New Urbanist wing—which retains a homogeneous flavor of past stylistic enthusiasms like neoclassicism, garden cities, art deco, and modernism. The importance of Smart Growth, and to a lesser extent New Urbanism, is its emphasis on the neighborhood scale, a broad range of objectives, and consensus building through local coalitions. These represent a fundamental paradigm shift away from the top-down, one-size-fits-all approaches of the past century.

As Smart Growth correctly advocates, we can no longer address such issues as transportation, housing, and environmental quality in isolation, that is to say through separate "stovepipe" agendas of particular bureaucracies and stakeholder interest groups. Decentralized grassroots urban advocacy involves creative blending and layering of diverse programs, laws, and funding sources—"like lasagna" one neighborhood activist calls it. As suggested earlier, Smart Growth has provided a segue from the Technocrat Decades to the Humane Decades. Hopefully, it will continue to help energize and mobilize local shirtsleeve efforts to make ordinary places more habitable, without prescribing what physical form those efforts must take. From here on, it is the *process* rather than the *product* that counts most.

Reinventing Affordable Housing

National efforts to provide decent public and subsidized housing for low and moderate-income families, as per the goal of the 1968 Housing and Urban Development Act, have long been frustrated by politics, bureaucracy, and inadequate resources. In 2002, the Millennial Housing Commission starkly declared: "Affordability is the single greatest housing challenge facing the nation. In 1999, one in nine households reported spending more than half its income on housing, while hundreds of thousands went homeless on any given night. Wide gaps also remain between the homeownership rates of whites and minorities, even among those with comparable incomes."[21]

The supply of affordable units in many cities has persistently declined due to demolition, obsolescence, lack of maintenance, or conversion to upscale condo or rental units, as in "Stuytown" in Manhattan. The disparity between average household income and average housing costs that I discussed in chapter 6 has continued to grow: among 46.2 million working households in 2009, nearly one-quarter paid more than half their incomes on housing, either ownership or rental.[22] This was an increase of 600,000 households from the 2008 number of households deemed to be severely burdened.[23]

The crisis in both public and affordable housing results from the complex interaction of many factors including the following four: (1) stagnation or decline
of income and net worth for most households other than the very wealthy;[24] (2)
the skewing of new home construction towards oversized and more costly homes;
(3) hostility to lower-cost or multifamily residential construction in many communities; and (4) rising costs of rental housing due to scarcity of available units.
The housing industry's "solution" to this has been to compound it through the
infamous practice of predatory lending based on the expectation that home prices
would rise. The failure of that outcome with the drop in home prices since 2006
has left more than ten million borrowers "underwater" (i.e., owing more than their
homes are now worth). Many of these borrowers have already lost their homes
or face the threat of foreclosure. Neighborhoods infested with vacant homes are
in turn dragged down financially, socially, and emotionally. This bleak record of
failed national housing policies nevertheless masks an underlying trend away
from centralized, technocratic approaches to a gradual emergence of decentralized, community- and consumer-based strategies.

Public Housing

Major changes in public housing strategies have ensued from public interest lawsuits and other forms of citizen protest against the status quo. I discussed in chapter 3 how the practice of warehousing poor African-Americans in high-rise public
housing projects was successfully challenged in the landmark case of *Gautreaux
v. Chicago Housing Authority.* The Federal District Court in 1969 held such segregated housing programs to be unconstitutional and ordered the housing authority
and the U.S. Department of Housing and Urban Development to diversify the
design, location, and occupancy of public housing.[25] Despite its tortuous history,
this case over time would help to revise national policies concerning public housing for the very poor.

In the spirit of *Gautreaux,* Congress adopted the "Section 8" housing voucher
program created by Congress in 1974. Under this program, low-income families and individuals who hold a Section 8 voucher may rent a dwelling unit in
the general housing market (if they can find one available) with their share of
the rent capped at 30 percent of their annual income and the remainder paid by
the U.S. Department of Housing and Urban Development (HUD).[26] Local housing authorities are authorized by HUD to issue a certain number of vouchers for
use within the geographic community that they serve. Although the *Gautreaux*
strategy sought to make Section 8 vouchers usable in suburbs as well as central
cities—a goal upheld by the U.S. Supreme Court in *Hills v. Gautreaux*[27] in 1976—
white communities long resisted Section 8 projects and tenants, with at least tacit
support from HUD and Republican administrations. But according to Alexander
Polikoff, the lead attorney in the case, "the longevity of the *Gautreaux* Program

gave 'housing mobility' a visible place on the housing policy landscape for over two decades. In 1987, Congress made Section 8 certificates 'portable' across housing authority jurisdictional lines. In 1991, Congress authorized a ten-year, five-city housing mobility demonstration program. [By 1998], some fifty-two mobility programs were operating around the country. About a dozen grew out of settlements of Gautreaux-type lawsuits against HUD, some were HUD programs that gave incentives to housing authorities to cooperate in regional Gautreaux-type arrangements, and a few were voluntary initiatives."[28]

Another *Gautreaux* legacy was the Housing Opportunities for People Everywhere Program (HOPE VI) signed by President George H. W. Bush in late 1992.[29] (The program was nearly abolished during his son's administration.) This initiative emerged from a finding by the National Commission on Severely Distressed Public Housing that some eighty-six thousand units of the nation's scarce supply of housing for the very poor were uninhabitable and countless others were poorly maintained and managed. HOPE VI virtually codified the *Gautreaux* judicial mandate by promoting small-scale new or renovated public units scattered among various neighborhoods and available for rent or sale to families and individuals of diverse race, age, and income status.[30] Moreover, consistent with the spirit of the 1990s, HOPE VI would embrace the basic design principles of New Urbanism (see fig. 7.1).

Over the course of fifteen years, HOPE VI grants to local housing authorities helped to demolish 96,200 public housing units and produce 107,800 new or renovated housing units, of which 56,800 were to be affordable to the lowest-income households.[31] The net loss of some forty thousand units was supposed to be offset by Section 8 vouchers issued to displaced tenants—a strategy impaired however by the shortage of both vouchers and units available for rent or purchase. Meanwhile, the private-sector redevelopment of projects like Cabrini-Green near downtown Chicago generated charges of gentrification—replacing housing for the poor with higher priced development for the middle class—an echo of Title I urban renewal of the 1950s and 1960s.[32]

Section 8 and HOPE VI, while steps in the right direction, have failed to adequately resolve the critical deficiency of housing units for the very poor. A *New York Times* editorial summarized the situation in late 2010: "After nearly two decades of weak financing from Congress, a large number of the public housing developments that shelter 2.3 million of the nation's poorest, most vulnerable are falling apart. . . . Today because of funding shortfalls, only one in four families that quality for federal rent support receive it. Families that do get to lease public housing units must often wait ten years or longer for the opportunity."[33]

Subsidized Housing

Affordable housing for low- to moderate-income households ("subsidized housing") has experienced a more pronounced break with former technocratic approaches.

In their book *Comeback Cities* (2000), Paul Grogan and Tony Proscio identify four trends with a "surprising convergence of positives" that underlay such shirtsleeve urbanism in the 1990s:[34]

1. The maturing of a huge, rapidly expanding grassroots revitalization movement.
2. The rebirth of functioning private markets in former wastelands.
3. The reduction in urban crime rates.
4. The unshackling of inner-city life from the giant bureaucracies that once dictated everything that happened there.

Unlike public housing which remains tethered to the HUD-Local Housing Agency model, subsidized housing programs typically operate through ad hoc partnerships involving some combination of government agencies, nongovernmental organizations, foundations, colleges and universities, and local community groups Such initiatives are often spearheaded by local neighborhood associations, community development corporations (CDCs), or faith-based programs like Habitat for Humanity.[35] The Dudley Street Neighborhood Initiative in the Roxbury/North Dorchester section of Boston has served as a national model in homegrown community redevelopment. Since it was formed in 1984, that project

Fig. 7.3. New affordable housing constructed by Dudley Street Neighborhood Initiative, Boston, ca. 2005.

(which serves a mixed population of white, African American, Latino, and Cape Verdeans), has converted hundreds of vacant lots from illegal dumps to building sites for several hundred units of affordable housing (fig. 7.3).[36]

With the onset of the civil rights movement in the 1960s, and within days after the assassination of the Reverend Martin Luther King Jr., Congress adopted the Civil Rights Act 1968 which banned discrimination on the basis of race, religion, and ethnicity (and later gender) in the private real estate industry.[37] Soon after that, the Housing and Urban Development Act of 1968[38] authorized subsidies to promote construction and occupancy of owned or rented homes for "low and moderate income" households.

"Redlining" of minority neighborhoods by mortgage lenders remained a major obstacle to construction, renovation, and purchase of homes and businesses in lower-income communities. Under pressure from housing activists, Congress adopted the Community Reinvestment Act of 1977 (CRA),[39] which requires lenders to treat all applicants and communities equitably within their service areas. Compliance with the act is monitored in periodic CRA audits by federal bank examiners. CRA has helped to release vast amounts of capital to borrowers in minority communities for use in construction, renovation, and purchase of structures, including affordable housing. CRA has been credited since its inception with turning around many distressed communities, prompting banks to channel more than $1 trillion (without a single dollar from Congress) into reinvestment projects.[40] The national research and action institute PolicyLink acknowledges that in addition to the $1 trillion community investment commitments, "the polices and practices of financial services have significantly changed; and the public's and media's views have shifted to support the need for equal access to capital and financial services-largely as a result of CRA advocacy."[41]

Congress also primed the pump of private investment in affordable housing through the Low Income Housing Tax Credit (LIHTC), first authorized in the Tax Reform Act of 1986. Under this provision, HUD allocates an annual quota of LIHTCs to each state. The states in turn distribute the credits to qualified developers of new or renovated affordable housing for rent or purchase. The developer then "sells" the credits to business corporations and other investors, using the funds so acquired to build the project. Investors apply the LIHTCs they have "purchased" to reduce their federal income tax liability over a ten-year period. The federal subsidy thus is the reduction in tax revenue due to the use of such credits. Grogan and Proscio attest the role of LIHTCs in broadening affordable housing development: "For all its restrictions (it's still a federal program, after all), the Tax Credit has given grassroots developers access to a source of publicly subsidized private capital that can be tailored to multiple needs."[42] By 2011, the LIHTC had stimulated the production or rehabilitation of some 2.4 million affordable homes since 1986.[43]

The primary source of direct federal funding for affordable housing has been the HOME Investment Partnership Program established by the George H. W. Bush administration in 1990. The program has since distributed $32 billion to state and local housing authorities, which in turn support local private-sector developers.[44] (However, a 2011 *Washington Post* investigation of 5,100 HOME projects then in progress nationwide found that nearly seven hundred projects—one in seven—have been stalled or abandoned, with inadequate oversight by HUD and local authorities.)[45]

Another acronymic device widely used to fund local community development and affordable housing is tax increment financing (TIF). This involves the designation of a special district by a local government encompassing a geographic area targeted for new commercial, industrial, or residential development. Any increase in property tax revenue that would result from future private investment is earmarked to defray the costs of site preparation and infrastructure needed to make the district more inviting to developers. Revenue bonds to be repaid with such future tax increments are issued to fund site improvements intended to lure such development. Chicago under Mayor Richard M. Daley (who served 1989–2011) embraced the TIF strategy robustly. More than 136 TIF districts in Chicago cover about thirty percent of the city's land area whose combined assessed value "increased by $11.4 billion, or nearly 16,000 percent between 1986 and 2005."[46]

The Atlanta BeltLine, one of the country's largest Smart Growth redevelopment projects combining affordable housing, light rail, bike trails, and community parks, relies heavily on TIF funding. A tax allocation district (TAD) was established in 2005 to earmark tax increments from the City of Atlanta, Fulton County, and Atlanta Public Schools, a district covering more than 6,500 acres of the city, which is projected in the next quarter-century to generate $1.7 billion in bonding capacity.[47] The BeltLine proposes to generate 5,600 units of affordable housing over twenty-five years. However, its first five years only yielded 120 units, and housing advocates are calling for leveraging available tax increment revenue with LIHTCs to provide more low-income units.[48]

Obviously, the earmarking of increased tax revenues *within* a TIF district to pay off the bond issue means that such funds are unavailable to enhance the community's general tax revenue. The broader community (including its schools) theoretically benefits from tax growth from spillover development outside the district. But this assumption has been challenged as unduly optimistic by researchers who found that growth *outside* of TIF districts in the Chicago area rarely offsets the revenue gain *inside* districts. To the contrary, assert the authors of one land policy study conducted in 2006, "commercial and industrial TIF districts both show a significantly negative impact on growth in commercial assessed values outside the district."[49]

Along with the evolution of flexible financial mechanisms like CRA, LIHTCs and TIFs has been the emergence of a generation of clever and agile nongovernmental housing providers that put such funding sources to use. Foremost among the myriad grassroots engines of local initiative is the Community Development Corporation (CDC). CDCs are incorporated nonprofit organizations that serve particular neighborhoods or communities through development of affordable housing and related facilities, economic development, and social services. CDCs specialize in assembling from diverse sources—government, foundation, LIHTCs, TIFs, Section 8, and so on—revenue that is used to accomplish their goals. By 2000, over 3,600 CDCs were operating in cities and neighborhoods across the country. Their collective efforts have been greatly facilitated by two national umbrella organizations formed in the 1980s to channel public and private money to CDCs: the Local Initiative Support Corporation (LISC),[50] launched by the Ford Foundation, and Enterprise Community Partners,[51] founded by the developer James Rouse and his wife. LISC as of 2010 had brokered $8.1 billion in Low Income Housing Tax Credits towards the creation of 277,000 affordable homes and apartments, housing 831,000 people.[52] Along with faith-based housing providers like Habitat for Humanity and other nonprofit developers, this sector now dominates the field of affordable housing in the United States.

According to CDC veteran Paul Grogan, writing in 2000, CDCs offered "an instructive counterpoint to the disappointing result of Washington's large-scale housing development programs of the 1960s and early 1970s. To begin with, CDCs aren't creatures of any single federal program, and thus aren't easily whipped about by Washington's annual bureaucratic intrigues and fiscal storms. They draw from many sources of money to accomplish multiple ends: Their purpose is not just to produce housing, but to produce housing as a catalytic and integrated element of overall community renewal."[53]

This assessment remains essentially valid even in the post-2006 era of the Great Recession, according to Joel Bookman, Director of Programs at Chicago LISC.[54] Among a dozen Chicago "new community" projects (recycling an old housing term) supported by his office, he suggested that I visit Logan Square on the northwest side of Chicago. Doing so in December 2011, I encountered a lively, bustling office of CDC staff and neighborhood volunteers, working amid walls covered with murals, posters, and photographs—a far cry from the dreary workplaces of HUD bureaucrats in Washington. Among their various projects, the Zapata Apartments when completed will provide sixty-one affordable apartments of varying size, and a community room, on four sites in the mixed-income neighborhood of Logan Square. The Zapata project is spearheaded by the Logan Square Neighborhood Association (LSNA), an active nonprofit community organization that turned fifty in 2012. LSNA is assisted financially and technically by LISC Chicago, which has designated Logan Square one of the sixteen "new community" projects it supports with funds from the MacArthur Foundation and other sources. (The mural on the back cover was photographed at Logan Square.)

The Zapata project results from intensive community discussions that identified a need for new residential development to serve the diverse population of Logan Square, in preference to higher-income newcomers ("gentrification"). Under LSNA oversight, and with predevelopment financing from LISC, the Zapata project will be constructed by the Bickerdike Redevelopment Corporation (BRC), itself a non-profit community-based entity founded in 1967, which has built or rehabilitated over 1,100 units of affordable housing in the area to date.

Funding for the project, largely secured at the time of this writing, exemplifies the "lasagna" approach to financing community development today: LIHTCs will provide 76 percent ($19 million) of the project's budget, TIF will provide 18 percent ($4.6 million), and other sources will provide the remaining 6 percent. The project also has contracted with the Chicago Housing Authority to provide ten two-bedroom units and eight three-bedroom units to low-income tenants with Section 8 vouchers. Groundbreaking for Zapata Apartments joyfully occurred on September 13, 2011.[55]

Americans with Disabilities

On July 26, 1990, President George H. W. Bush signed one of the most deeply humanitarian acts in the nation's history: the Americans with Disabilities Act or "ADA."[56] Remarkably, in light of our paralytic politics two decades later, the U.S. Senate adopted the law by a vote of 76–8 and the House passed it by unanimous voice vote.

ADA greatly expanded the scope of federal protection against discrimination beyond the categories of race, religion, national origin, and gender addressed in earlier civil rights laws to embrace persons afflicted by "a physical or mental impairment that substantially limits a major life activity,"[57] Specifically, the act's preamble states: "Discrimination against individuals with disabilities persists in such critical areas as employment, housing, public accommodations, education, transportation, communication, recreation, institutionalization, health services, voting, and access to public services."[58]

On behalf of all persons entitled to its protection, the act broadly mandates remedial actions by state and local governments, businesses, institutions, and organizations (other than religious bodies and private clubs) to alleviate discrimination in employment, public services, public accommodations, and telecommunications. The act is administered and enforced by the federal Department of Transportation and the Department of Justice.

Among ADA's most widespread and significant achievements has been the removal or reduction of physical barriers to mobility in public buildings, hotels, offices, stores, theaters, libraries, museums, parks, hospitals, public transportation, and other facilities accessible to the public in general. Moreover, the act required not only that new construction be ADA-compliant but that existing facilities be retrofitted to be accessible, subject to certain economic criteria with the costs of compliance to be covered by the facility's owner, not the taxpayer.

ADA access requirements have not only assisted people with disabilities, but have in various ways made our shared spaces more amenable for the general public. Elevators and wheelchair ramps have been added to countless older buildings whose steep stairways may be challenging to anyone with short legs or a heavy load to carry. Ubiquitous sidewalk curb-cuts at street intersections benefit not only those in wheel chairs but also parents pushing strollers, cyclists, and people pulling roll-along suitcases. Public buses "kneel," children love opening doors with push buttons, rest rooms include spacious cubicles, and water fountains are within the reach of kids and other height-challenged people. Indeed, ADA has the distinction of being an unfunded federal mandate that benefits just about everyone.

Green Buildings

In tandem with Smart Growth and the ADA, the green building movement—another product of the 1990s—promotes more humane and sustainable buildings, streetscapes, and neighborhoods. The movement has already made a substantial impact on the design of the nation's built environment through the advocacy and professional standards established by the U.S. Green Building Council (USGBC), an influential nongovernmental organization established in 1993.

The green building movement is largely market-based, operating through voluntary professional practice standards that enhance the value and reduce costs of building construction and operation. Although using green building methods is not directly mandated by federal law as with the ADA, the movement has rapidly gained political influence as reflected in a plethora of state and local policies that encourage and sometimes require the adoption of green building technology. USGBC in 2011 encompassed 79 local affiliates, nearly 16,000 professional, trade, and research member organizations, and more than 174,000 certified practitioners.[59] Internationally, the World Green Building Council embraces some 90 national green building councils that promote green building practices in their respective countries.

At the heart of USGBC's strategy is its famous Leadership in Energy and Environmental Design (LEED) system for certifying green construction practices. LEED aims to provide building owners and operators with a framework for identifying and implementing practical and measurable green building design, construction, operations and maintenance solutions, and to promote "sustainable building and development practices through a suite of rating systems that recognize projects that implement strategies for better environmental and health performance."[60]

LEED establishes measurable criteria for five primary variables in the siting and construction of new buildings or renovation of existing structures:

1. *Sustainable Sites*—discourages development on previously undeveloped land; seeks to minimize a building's impact on ecosystems and waterways; encourages regionally appropriate landscaping; rewards smart transportation choices; controls stormwater runoff; and promotes reduction of erosion, light pollution, heat island effect, and construction-related pollution.

2. *Water Efficiency*—encourages smarter use of water, inside and out. Water reduction is typically achieved through more efficient appliances, fixtures and fittings (inside) and water-conscious landscaping (outside).

3. *Energy & Atmosphere*—encourages a wide variety of energy-wise strategies such as efficient design and construction; efficient appliances, systems and lighting; and the use of renewable and clean sources of energy, generated on-site or off-site.

4. *Materials & Resources*—encourages the selection of sustainably grown, harvested, produced, and transported products and materials; promotes waste reduction as well as reuse and recycling; and particularly rewards the reduction of waste at a product's source.

5. *Indoor Environmental Quality*—promotes strategies that improve indoor air as well as those that provide access to natural daylight and views and improve acoustics.[61]

In addition to the five points above, LEED offers credits relating to certain less technical variables: locations and linkages; awareness and education; innovation in design; and regional priority.

Certification of practitioners and projects is fundamental to the operation of the LEED rating system. The Green Building Certification Institute (GBCI), established as an independent entity in 2008, administers the certification process. Projects submitted for evaluation by GBC are granted points reflecting how well they meet the various criteria established for each of the above factors. The total of all points determines whether the project qualifies for one of LEED's four levels of merit: Certified, Silver, Gold, or Platinum. LEED certification confers "bragging rights" on developers, owners, and tenants of commercial, residential, institutional, and other projects. At this writing, USGBC estimates that 1.6 billion square feet of new and renovated commercial space has been certified.

As with disability access under ADA, the green building movement is changing the entire culture of the building industry. It promotes energy efficiency, water conservation, recycling of building materials, and comfort for building occupants. Such simple measures as enhancing natural daylight inside school buildings, for instance, are found to improve student and teacher morale and reduce energy consumption. "Green roofs" or rooftop gardens have sprouted on hundreds of old and

Fig. 7.4. Mayor Richard M. Daley's signature green roof atop Chicago's City Hall (the far side of the roof, above the Cook County half of the building, remains asphalt).

new buildings, most famously on top of the Chicago City Hall (fig. 7.4).[62] Walkability, bike lanes, tree-shaded streets, and proximity to public transit—all Smart Growth articles of faith—are rewarded in the LEED certification process. Recently, USGBC, the Congress for the New Urbanism, and the Natural Resources Defense Council have jointly developed criteria for a new LEED category, Green Neighborhood Development.[63]

Recreational Trails and Greenways

A month before leaving office, President Bush on December 18, 1991, signed another major catalyst for humane urbanism awkwardly named the Intermodal Surface Transportation Efficiency Act (ISTEA). While much of its six-year, multi-billion-dollar budget was earmarked to complete the Interstate Highway System, ISTEA for the first time authorized a portion of the federal Highway Trust Fund to be devoted to recreational and urban trails.

The modern trails movement in the United States dates back to the (literally) path-breaking proposal for the Appalachian Trail in 1921[64] by Benton MacKaye, which he later refined in his 1928 book *The New Exploration.*[65] As a landscape visionary and charter member of Lewis Mumford's Regional Planning Association of America, MacKaye thought of the trail as a "levee" or bulwark to resist the sprawl of the nation's fast-urbanizing eastern seaboard. The trail itself became the famous 2,100-mile hiking route connecting a vast chain of public parks extending along the Appalachian highlands from Georgia to Maine.

The Appalachian Trail inspired Laurance S. Rockefeller's Outdoor Recreation Resources Review Commission (ORRRC), discussed in chapter 5, to propose a national system of long-distance trails through America's most scenic natural areas, culminating in the National Trails System Act of 1968.[66] That law and its successors have established today's network of twenty national trails (eight scenic trails and twelve historic trails) extending almost forty thousand miles.[67]

In tandem with the long-distance national trail system, the idea to reuse abandoned rail rights-of-way for bike and pedestrian paths ("rail trails") originated with a letter to the *Chicago Tribune* of September 25, 1963, from the naturalist May Theilgaard Watts.[68] The letter praised the English tradition of public footpaths and pointed out the opportunity to establish a pedestrian and bike path along a twenty-seven-mile abandoned rail right-of-way through Chicago's older suburbs west of the city.

Watts's proposal generated a wave of interest among the towns and grassroots leaders along the proposed route. In 1964, the Illinois Prairie Path (IPP) incorporated as a nonprofit advocacy group to promote the idea. With a short film created with the help of Chicago's Open Lands Project, IPP generated growing community interest and support from local governments and corporations sharing the right-of-way. Assembly of the trail route began with purchase of a major segment by DuPage County that was leased to IPP. Additional segments were obtained under leases or easements to IPP from utility companies and other landowners. While IPP manages the overall trail, most of the path has been "adopted" by local civic and garden clubs and Scout groups for purposes of routine maintenance.[69] Since the 1960s, the Illinois Prairie Path has expanded to a total of sixty-one miles extending from Maywood near Chicago some twenty miles to Wheaton and then dividing into branches connecting to various communities along the Fox River valley (formerly the western edge of metropolitan Chicago).

The model of the Illinois Prairie Path helped to stimulate similar efforts elsewhere. In 1986, with some two hundred rail trails in various stages, the national Rails to Trails Conservancy was established. Five years later the conservancy, already with a membership of forty thousand, helped to secure adoption of ISTEA in 1991. This law and its successors in 1998 and 2003 have significantly assisted states, localities, and trail advocacy organizations in planning, developing, and maintaining a nationwide network of trails in and near metropolitan areas. Total federal funding toward recreational trails (rural and urban) in 2010 amounted to about $50 million. Rail trails now amount to nearly twenty thousand miles—a widely distributed system of off-road corridors for walking, jogging, cycling, in-line skating, cross-country skiing, and other recreational uses.[70] These in turn are only a small proportion of a much larger network of foot and bike paths of many kinds that interlace the nation and its metropolitan areas (fig. 7.5). (I discuss New York City's new High Line in chapter 8 and the "Walkway over the Hudson," in chapter 10.)

Even with rail trails however, technocracy lurks like Japanese knotweed, ready to spring to life wherever federal money is disbursed. In my community of Northampton, Massachusetts, our original rail trail built with local and state funds in the 1980s is a serene and scenic bikeway that blends into its surrounding landscapes of forest, meadow, wetland, or park. Newer segments of our regional rail trail system built in the post-ISTEA era are conspicuously over-designed by highway engineers: grim retaining walls topped with boring arborvitae shrubs, heavy wooden fencing, and superfluous, standard-issue signage—in short the federal standards drive up the cost per mile (and profit to contractors) while walling off surrounding landscapes behind protective structures and jungles of invasives (such as knotweed and sumac) that smother wildflowers and other local flora.

Rail trails and "water trails" for nonmotorized enjoyment of local streams and rivers often serve as the sinews of much larger multipurpose greenways. Charles E. Little in the mid-1990s described several regional greenway systems then (and still) in progress involving some combination of urban river greenways, pedestrian and cycling trails, ecological corridors, and scenic drives and historic routes.[71] The "grandmother of regional greenways" certainly was the chain of greenspaces in Boston known as the Emerald Necklace, designed by the firm of Frederick Law Olmsted and Calvert Vaux in the 1880s to connect Boston Common and the Public Garden to the Fenway, Jamaica Pond, Arnold Arboretum, and Franklin Park. The popularity of this "necklace" in turn inspired proposals as early as the 1890s for a wider Bay Circuit of regional parks encircling suburban Boston. While much of the proposed route was later preempted by the Route 128 beltway and associated development, many elements of the Bay Circuit have been protected as state or metropolitan parks, or properties acquired by The Trustees of Reservations, Massachusetts Audubon Society, and local conservation commissions.

Elsewhere, regional greenways vary from narrow bike and pedestrian corridors through urban areas like the Brooklyn-Queens greenway and its counterpart, the Manhattan Waterfront Greenway,[72] to broader planning regions such as the Hudson River Valley Greenway. The latter, spearheaded by Scenic Hudson, Inc. and the Hudson River Valley Greenway Council established by the state in 1988, has assembled both land and water-based "trails" between New York City and Albany as the spine of a much larger planning region extending to the state line to the east, up to fifty miles west of the river, and north to the Adirondacks.[73]

Fig. 7.5. Former rail bridge over the Connecticut River at Northampton, Massachusetts, now a busy link in the regional bikeway system.

Massachusetts in 1995 established its Connecticut River Greenway State Park, a network of state parks and boat launch sites lining the river within the state's borders.[74] The San Francisco Bay Trail, much expanded since Charles Little's mid-1990s case study, today includes over 310 miles of completed segments out of an eventual goal of five hundred miles encircling the entire bay shoreline.

Urban Trees and Forests

Urban forestry is one of the most widespread and visible elements of humane urbanism, as reflected in the "Million Tree" planting programs in New York, Chicago, and Los Angeles and their smaller counterparts in other cities. (See figs. 10.3 and 10.5 in chapter 10.) The current rage for urban trees and forests is the latest phase of a long and conflicted relationship between urban dwellers and trees in their vicinity. Colonial settlements regarded available woodlands, in the tradition of the English commons, as sources of firewood, building materials, masts and tar for ships, and sustenance in the form of nuts, berries, and fruits. The aesthetic values of trees in the built environment gained a limited start with planned cities such as Philadelphia and Savannah endowed with English-style squares. Otherwise, trees in or near cities were generally treated as utilitarian elements of nature's bounty just like water, soil, and rock.

With the advent of the urban parks movement in the mid-nineteenth century, the pioneering landscape architects Andrew Jackson Downing and Frederick Law Olmsted popularized the appeal of trees as ornamental features in the urban environment. Modeled on Joseph Paxton's Birkenhead Park near Liverpool, the 1858 "Greensward" Plan designed by Olmsted and Vaux's firm for New York's Central Park juxtaposed groves of woodlands and freestanding trees among grassy meadows, formal gardens, carriage roads, bridle paths, and water features. This was not an exercise in preservation of a native landscape like the Indiana Dunes. Rather the trees in the Greensward Plan and later Olmsted-firm works like Brooklyn's Prospect Park and Philadelphia's Fairmont Park were selected, planted, and maintained to create a vivid and ever-changing palette of blossoms and foliage throughout the growing season. Olmsted's landscape plan for the 1893 World's Columbian Exposition in Chicago and Georges Haussmann's tree-lined Paris boulevards helped to establish decorative arborculture as a standard element of City Beautiful urban plans.

Meanwhile, European urban forests like the Vienna Woods inspired similar efforts to protect existing stands of natural forest within or near certain American cities. Establishment of Greater Boston's Metropolitan Park Commission in 1892 and the Cook County Forest Preserve District in 1913 signaled a new trend toward preserving forests and other natural areas in their natural state rather than designing and planting them to be "decorative." The primary purpose of such forest reservations was to provide rest and recreation to city or suburban families able to reach them by streetcar, commuter rail, and later automobile. Competition has often pitted natural resource conservation against demands for roads, parking lots, concessions, and commercial recreation facilities. Over time, many forest preserves and city park trees have suffered from overuse, neglect of maintenance, soil compaction, hydrologic disturbance, insect infestations, fires, and strangulation by invasives such as bittersweet and honeysuckle vines.

The Technocrat Decades, needless to say, were disastrous for urban trees and forests. Urban renewal, highway construction, and sprawl took little notice of trees in their path. Urban parks were so commonly ravaged for highway corridors that Congress in the 1966 Highway Act was moved to enact Section 4(f) which required consideration of alternate routes other than traversing a park (but the alternative might instead destroy settled neighborhoods, an even less desirable option). Jane Jacobs's successful crusade to save Washington Square and its trees from a Robert Moses road project was the best known of many battles nationwide on behalf of urban trees, forests, and parks in the path of highways and sprawl. (I discus the saving of Thorn Creek Woods near Chicago in the early 1970s in a box in chapter 5.)

With the onset of the environmental movement and especially the adoption of the federal Clean Air Act in 1970 and its state counterparts, urban trees and forests gained new significance as contributing to air quality improvement and moderation of local temperature extremes. Research by the U.S. Forest Service documented that strategically located shade trees help to mitigate the "urban heat island" that raises local temperatures and air conditioning costs due to the replacement of natural land cover with paved surfaces and rooftops.[75] The root systems of trees also help to stabilize soil erosion on steep slopes; the loss of trees in wildfires often is followed by mudslides in urbanized western mountain canyons. Tree cover and natural land surfaces also help to reduce storm runoff and thereby reduce pollution of local water bodies from "combined sewer overflows" (CSOs), a chronic issue for older cities where stormwater overwhelms local treatment capacity.

Los Angeles, one of the most smog-bound cities in the nation, was appropriately among the first to embrace urban tree planting as an air pollution mitigation strategy. In 1981, the city adopted an air quality management plan calling for the planting of one million trees over the next twenty years at a cost of $200 million. That goal was in fact largely accomplished within three years in time for the 1984 Los Angeles Olympic Games through a Herculean effort by city agencies, nonprofit organizations, and volunteers. This initiative was spearheaded by TreePeople, a nongovernmental organization founded by Andy Lipkis in 1978, which would become a respected advocate for urban tree planting and urban environmental improvement around the nation and globe.[76]

In the 1980s, urban trees and forests were found to provide emotional benefits for hospital patients with views of natural scenes from their windows who recovered more quickly than comparable patients without such views.[77] The psychological values of trees for urban residents were further explored by the U.S. Forest Service and the Morton Arboretum, who documented that people form "deep emotional ties" with trees associated with their childhood, family members, or special places in their lives. The planting of memorial trees is one cultural manifestation of that attachment. Furthermore, trees were found to have a calming effect on urban dwellers encountering them in preserves like botanical gardens.[78] Even great landmark trees that survive in ordinary neighborhoods may be a source of well-being and comfort for people having daily awareness of them. (See the section titled "A Tree Grows in Poughkeepsie" in chapter 10.)

In the 1990s, the refinement of geographic information systems (GIS) and satellite remote sensing capabilities enabled researchers to analyze urban tree canopy and estimate its fluctuations over time. American Forests, a nongovernmental organization based in Washington, DC, estimated losses of tree cover for various cities using its proprietary software package known as "CITYgreen."[79] In 1999, the *Washington Post* featured CITYgreen images on its front page revealing a 64 percent decline in the District of Columbia's tree canopy between 1973 and 1997. The article and its graphics inspired local resident Betty Brown Casey to establish a nonprofit organization, Casey Trees, Inc., to assist the city in planting trees and educating the public about their value and care.[80]

The city of Chicago under Mayor Richard M. Daley launched a 500,000 tree planting program in 1990 which was achieved by 2005, along with other initiatives including green roofs, LEED standards for public buildings, green streets and alleys (see fig. 7.4). By then, growing concerns about global warming highlighted yet another function of urban trees and forests: carbon dioxide sequestration, along with reduction of ground level ozone and heat island effects. Chicago's 2007 Climate Change Action Plan and 2009 Urban Forest Initiative committed the city to add another million trees by 2020.[81]

Los Angeles in 2006 under Mayor Antonio Villaraigosa embarked on its second Million Tree planting campaign with TreePeople again playing a key role. New York City has passed the halfway point in its own Million Tree program, a major element of Mayor Michael Bloomberg's *PlaNYC*—the city's 2030 sustainability plan. Counterpart tree planting programs are in progress in smaller cities across the country, combining the recovery of ecological services provided by urban forests with training in arborculture and the creation of green jobs for young people in distressed urban communities. (See case studies of New Haven and Marvin Gaye Greenway in chapter 10.)

Managing Urban Water Resources

Sprawl disturbs and sometimes devastates metropolitan water resources. Some of its impacts, as discussed in the previous chapter, include surface and ground water pollution, sedimentation, ecological degradation, wetland destruction, aggravation of flooding, and impairment of recreational uses. Among a plethora of federal water legislation since the 1960s, two were specific responses to horrific water-related disasters: the fire on Cleveland's Cuyahoga River in June 1969, that helped build support for the landmark 1972 Federal Water Pollution Control Act Amendments,[82] and the Love Canal toxic waste disaster in Niagara Falls, New York, in the 1970s that led to the federal "Superfund" Act of 1980[83] and its state counterparts. Likewise, the National Flood Insurance Act of 1968 and its amendments were adopted in the wake of a series of coastal hurricanes and inland floods that devastated homes and businesses built unwisely in the path of Mother Nature.

This is not the place to wade into the quagmire of federal water law. Suffice it to say that Congress has fortunately provided some key tools (table 7.1) for rehabilitating urban streams, wetlands, waterfronts, and watersheds, which have stimulated local initiatives such as those described in chapter 9. The balance of this section will briefly highlight three federal contributions to water resource management in metropolitan areas: water quality, wetlands protection, and floodplain management.

Urban Water Quality

Perhaps the most significant element of the 234-page federal Clean Water Act (as of 2011) and its volumes of regulations is the National Pollutant Discharge Elimination System (NPDES) established by the 1972 Water Pollution Control Act. That

7.1 Selected Federal Water Laws: 1960s–1980s

1965	Water Quality Act (PL 89-234)
	Water Resources Planning Act (PL 89-80)
1968	National Flood Insurance Act (PL 90-448, Title XIII)
1970	National Environmental Policy Act (NEPA) (PL 90-190)
1972	Federal Water Pollution Control Act Amendments (PL 92-500)
	Coastal Zone Management Act (PL 92-583)
1973	Flood Disaster Protection Act (PL 93-234)
	Endangered Species Act (ESA) (PL 93-205)
1974	Safe Drinking Water Act (SDWA) (PL 93-523)
1977	Clean Water Act (CWA) (PL 95-217)
1980	Comprehensive Environmental Response Compensation and Liability Act (CERCLA; "Superfund") (PL 96-510)
1986	Superfund Amendments and Reauthorization Act (SARA) (PL 99-499)
1987	Clean Water Act Amendments (PL 100-4)

landmark statute prohibited "point source" ("end of pipe") discharge of wastes into "waters of the U.S." without a permit. As administered by the U.S. Environmental Protection Agency established in 1970, the NPDES sought to eliminate such discharges over time. Control of toxic pollutants was further regulated under the 1977 water quality amendments. The NPDES over the past four decades has drastically reduced waste discharges from industrial sources and municipal sewage treatment plants, with corresponding improvement of water quality in many of the nation's rivers, lakes, and harbors. The 1987 water quality amendments further expanded federal regulation to include "nonpoint" or diffuse pollution sources including "land runoff, precipitation, atmospheric deposition, drainage, seepage or hydrologic modification."[84]

Older cities commonly constructed sewer systems that received both household wastes and storm runoff. During intense rainfall events, the combined flow exceeds the capacity of local sewage treatment plants and must be discharged untreated into available rivers, lakes, or harbors. Such "combined sewer overflows" (CSOs) long defied control under the NPDES system due to the immense cost and disruption of building separate collection systems for storm and "sanitary" sewage. In 1994, the EPA issued a new CSO policy designed to phase out such pollution sources over time. The policy involved a series of short-term management practices to minimize the impacts of CSOs while requiring that cities develop Long-Term Control Plans to eliminate the problem. Several cities like Chicago and Milwaukee have constructed tunnel and reservoir projects (TARPs) to store stormwater. An important strategy to control storm runoff has been the Smart Growth approach of low-impact development (LID) including rain gardens, green roofs, pervious pavement, and other landscape elements that help to retain precipitation onsite rather than discharging it to local storm sewers.

Toxic wastes discharged or stored at industrial sites across the country pose a more sinister and intractable threat to public health. Leaking storage drums containing poisonous substances became linked to local clusters of cancer and other illnesses at industrial sites like the W. R. Grace chemical plant in Woburn, Massachusetts, publicized by Jonathan Harr's 1995 book *A Civil Action* and the film based on it. Fundamental to stoking public outrage and government response was the realization that toxics discharged on or beneath the land surface seep into local surface and ground water, thus endangering the health of anyone who drinks water from such polluted sources.

Love Canal in Niagara Falls, New York, was the scene of the nation's most notorious and influential groundwater contamination disaster. Love Canal was an artificial ditch used for decades as a waste dump by the Hooker Chemical Company. In 1977, families living in the vicinity of the canal (and likely working at the plant) began to notice a rash of medical red flags (birth defects, high rates of cancer) and foul-smelling chemical wastes oozing into basements. After denial by Hooker, city, county, state, and federal officials conducted an investigation. Based on preliminary findings, President Jimmy Carter issued a "major disaster" declaration for Love Canal in 1978, the first such declaration for a technological disaster. With federal disaster assistance thus authorized, the state acquired the homes and assisted the relocation of 237 families from the vicinity. The canal was sealed off and a remedial drainage project initiated. Subsequently, it became apparent that the chemicals had contaminated groundwater more widely than originally estimated. And the brew of toxic chemicals was found to contain, in addition to benzene and a dozen other carcinogens, a measurable amount of the most deadly chemical ever synthesized: dioxin. The Love Canal Homeowners Association campaigned for further public assistance. Eventually, more than six hundred homes were acquired.[85]

In an enlightened response to Love Canal, Congress in 1980 adopted the Comprehensive Environmental Response Compensation and Liability Act ("Superfund"). The act charged the U.S. Environmental Protection Agency with establishing a National Priories List of known toxic waste sites and cleaning up the worst of them, with the costs to be paid by the responsible party when they could be identified. The tortuous subsequent history of Superfund and its state counterparts is beyond the scope of this discussion, but over time these programs have gradually remediated dozens of urban sites, allowing them to be reused for parks, housing, and other purposes.

While most Superfund projects have involved land sites, there have also been protracted battles over grossly polluted rivers and other urban waterways. When a specific corporate offender can be identified, the strategy for dealing with the problem is subject to prolonged debate, as with General Electric's polychlorinated biphenyl (PCB) discharges into the Hudson and Housatonic rivers. (The company argues that dredging, at its expense, will be more hazardous than allowing the toxins to remain buried under layers of sediments.) Often the parties responsible for toxic waterway pollution are unknown, too numerous, or no longer in business, leaving the clean-up costs to be borne by taxpayers. The selection of remediation strategy and its impacts on the local community add additional complexity. At this writing, EPA faces a "legal nightmare" in its efforts to clean up portions of the Passaic River in northern New Jersey.[86] And its choice of aeration to oxygenate the notoriously polluted Newtown Creek in Brooklyn is thought to be spreading infectious bacteria into the local atmosphere.[87]

Urban Wetlands

Turning to the slippery topic of "wetlands": The term itself embraces a wide spectrum of hydrologic features generally defined as "areas that are inundated or saturated by surface or ground water at a frequency and duration sufficient to support, and that under normal circumstances do support, a prevalence of vegetations typically adapted for life in saturated soils. Wetlands generally include swamps, marshes, bogs, and similar areas."[88]

Wetlands are as diverse as the vast grasslands of the everglades, isolated mountain swamps, and "prairie potholes" of the Great Plains. Depending on their regional environment, metropolitan areas are sprinkled with coastal salt marshes, freshwater swamps and bogs, and riparian wetlands along streams, lakes, and ponds. In their natural state, certain types of wetlands provide "ecological services" such as flood reduction, water purification, plant and wildlife habitat, and visual amenity. Until the 1960s, wetlands were regarded as waste areas, to be filled for agriculture or buildings, or used as receptacles for urban trash and industrial waste. (Fresh Kills on Staten Island, once a vast wetland, became New York City's primary trash disposal site—including debris from the World Trade Center—until closed and replanned for an eventual regional park.)

In 1969, the naturalists John and Mildred Teal published a small book, *Life and Death of the Salt Marsh*,[89] which, among other natural history writings, helped to galvanize public interest in stemming the further loss of those and other wetlands. Certain states and communities began to regulate wetlands alteration with the support of state court decisions in New Jersey,[90] Massachusetts,[91] Wisconsin,[92] and elsewhere. But in the absence of federal intervention, metropolitan sprawl continued to devour wetlands widely.

Ironically, Congress in the 1972 Water Quality Act laid a foundation for federal wetlands regulation without even intending to do so. The law nowhere mentions the term "wetlands." An obscure provision, Section 404, merely prohibits "the discharge of dredge or fill material into the navigable waters"[93] without a permit from the Army Corps of Engineers—a virtual restatement of federal policy dating back to the 1899 Rivers and Harbors Act. The Army Corps accordingly interpreted Section 404 to limit its permit authority over dredge and fill activities to "navigable waterways" and nowhere else. The Natural Resources Defense Council, on behalf of the ascendant environmental movement of the time, challenged this literal reading of the law. In 1975, a federal court[94] held that Congress really intended (despite the wording of Section 404) for federal permitting authority to extend broadly to "Waters of the United States" (a phrase used elsewhere in the 1972 act). As a result, new rules issued jointly by the Army Corps and EPA in 1978 went to the opposite extreme, proclaiming federal regulatory jurisdiction over dredge and fill activities in "navigable waters of the United States, construed very broadly.[95]

George H. W. Bush, in January 1989, declared a new national goal of "no net loss" of wetlands, meaning that any permitted disturbance of wetlands should be mitigated by at least equivalent acreage and type of wetlands being created or restored. This contributed to the practice of "wetlands banking" whereby an Army Corps permit might require the recipient (e.g., a highway department or shopping center developer) to pay for the preservation, creation, or restoration of wetlands of equivalent size and type somewhere else in the region or state. "No net loss" has remained at least a rhetorical goal under the Clinton, George W. Bush, and Obama administrations, but its realization in practice has been elusive.

In addition to its Section 404 permitting role, the Army Corps of Engineers has become, when so directed by Congress, a born-again wetlands restoration agency. This role began with the Charles River watershed in metropolitan Boston (described in chapter 9). Thus, the Army Corps, in its traditional technocratic role as builder of flood control and navigation projects, has been tempered by its "greener" regulatory and wetlands protection activities, which contribute (albeit unevenly and often contentiously) to grassroots efforts to save and restore remaining patches of urban wetlands.[96]

Flood Hazards

Finally, the nation's policies regarding flood hazards along urban waterways have evolved from reliance on "structural" flood control projects to "nonstructural" measures to manage land use and building in hazardous locations. The National Flood Insurance Program (NFIP) has been the mainstay of federal response to floods since it was established in 1968. Pursuant to an expert review of federal flood policies,[97] the NFIP employs several nonstructural strategies to reduce flood losses including floodplain mapping, floodplain management, and flood insurance, and to a modest extent land acquisition in areas chronically threatened by floods.

The purposes of the NFIP are two-fold: to shift the costs of flood losses from tax-based disaster relief to premium-based flood insurance,[98] and to reduce future flood losses through land use and building regulations. As the flood policy review panel observed, the two goals are interrelated: unless future construction in floodplains is restrained, the NFIP may inadvertently encourage new growth in hazardous areas. But the idea of federal control over land use in floodplains has always been politically unviable. Therefore, a clever "back-door approach" was adopted: the NFIP establishes minimum standards for local communities to adopt as a prerequisite to the availability of flood insurance within a given community. As with regulation of wetlands (which often overlap geographically with floodplains), the NFIP was initially stymied by a lack of maps indicating where local regulations should be applied. To address that need, the NFIP undertook a massive effort to map the nation's floodplains at a scale suitable for local land use planning and regulation. Over several decades, the program has spent billions of dollars on mapping "regulatory floodplains" (i.e., subject to a 1 percent risk of flooding in any year) bordering rivers, lakes, and coastal waters nationwide.

Virtually every urban and suburban community now participates in the NFIP. But as of the time writing, the NFIP is substantially in debt to the U.S. Treasury for payments far exceeding its premium revenue relating to Hurricane Katrina in 2005 and Superstorm Sandy in November 2012.[99]

Endangered Species

The Endangered Species Act of 1973 (ESA)[100] is the cornerstone of federal efforts to protect endangered or threatened flora and fauna. Some have considered it "the nation's toughest environmental law, a measure so strict that it can stop a $100 million dam project to protect a rare fish or ban logging on millions of acres of federal land to save an owl."[101] Many critics of the have often have assailed it as rigid and inflexible. The listing of a species as "endangered" and the preparation of a "species recovery plan" are based on scientific evaluation by the U.S. Fish and Wildlife Service (FWS) that is often the subject of heated dispute.[102]

ESA not only protects endangered plants and wildlife but also seeks "to provide a means whereby the ecosystems upon which endangered species and threatened species depend may be conserved."[103] This implies that land use planning and regulation will be utilized to conserve ecosystems and the species that depend upon them. Controversially, this requirement applies to privately owned land as well as public land. Disputed applications of the act have included limitations on logging of old-growth forests in the Pacific Northwest to protect the northern spotted owl, the temporary closing of public bathing beaches along the Atlantic coast to protect nesting habitat for the piping plover, and proposed limits on development in the Florida Keys to protect the diminutive Key deer and the Stephens' kangaroo rat in western Riverside County, California.[104]

In 1982, Congress amended the ESA to permit limited "takes" of designated habitat and species pursuant to the development of a habitat conservation plan (HCP) for a particular area. An HCP requires a formal planning process involving all interested parties (e.g., landowners, developers, local governments, state and federal wildlife agencies, and environmental organizations). This process seeks to achieve an agreement among all parties that specifies three areas of concern: (1) the impacts which will result from proposed land use changes; (2) steps to be taken to minimize and mitigate such impacts and funding to implement those steps; and (3) alternatives to the "taking" and why they were not adopted. HCPs have been used most frequently in the Southwest to allow land developers authority to encroach upon sensitive lands ("incidental take") in exchange for donations of land or other measures to otherwise protect list species and habitats.

The Riverside–San Bernardino metropolitan area east of Los Angeles grew in population by 25.7 percent during the 1990s and was deemed to be the nation's most sprawling metro area by Smart Growth America.[105] The ecologically diverse region also supports a substantial proportion of all threatened or endangered species in California. In 1998, Riverside County launched an ambitious regional plan process to coordinate habitat, land use, and transportation known as the Integrated Project. A major component of the project is the Multiple Species Habitat Conservation Plan (MSHCP), which calls for public acquisition of 153,000 acres for additional wildlife habitat. Remaining undeveloped private land in western Riverside County is subject to wildlife protection conditions to be administered through narrative criteria for 160-acre "cells." According to the Endangered Habitats League, the plan balances new housing and highways with "an ambitious conservation plan whose enactment in the political climate of the Inland Empire (as the metro region is called) is little short of miraculous."[106] The foreclosure crisis and concomitant collapse in new construction in Riverside County since 2006 has left the plan in limbo, while easing immediate pressure on the region's remaining ecological habitat, at least temporarily.

Chapter 8

New Age "Central Parks:" Two Grand Slams and a Single

The greatest city parks . . . often become the very symbols of their cities, the central touchstones of memory and experience for residents and tourists alike.
—PETER HARNIK, *Urban Green*, 2010

Millennium Park is a true commons. People of every age and ethnicity un-self-consciously merge and mix. The location is as public as public gets.
—ROBERT CAMPBELL, "Chicago Hits This One Out of the Park"
Boston Globe, 2005

Older American cities are richly endowed—even cities that themselves are no longer rich—with parks long ago established by visionary civic leaders, local philanthropists, and creative park designers. In 1634, the Boston Common was established by the earliest Puritan settlers for grazing of livestock, training militia, outdoor exercise, and burial of the dead, and the "hanging of unwelcome Quakers."[1] Based on English precedents, colonial town commons or greens later became treasured public parks at the core of many New England cities and towns, today serving as venues for community festivals, sports, farmers markets, political gatherings, and the parking of vehicles. Elsewhere, other city parks originated as military garrisons like the Presidio in San Francisco, New York's Battery Park and Governors Island, and the Battery in Charleston, South Carolina.

A new age of park creation "from the ground up" began with New York's Central Park in the mid-nineteenth century, one of the world's best-known and beloved urban parks. Central Park was not anticipated in the city's 1811 Commissioners Plan, which projected a grid of streets and real estate development relentlessly marching up the length of Manhattan Island. The 840-acre site of the park was snatched from this fate through its purchase by the city in response to the public advocacy of the landscape architect Andrew Jackson Downing and the newspaper editor and poet William Cullen Bryant. In 1858, the city selected the famous "Greensward Plan" for the new park co-designed by Frederick Law Olmsted and Calvert Vaux, who oversaw the transformation of the rugged site into an artificial landscape of meadows, wooded glades, great lawns, gardens, and water features. Although Central Park included bridle paths to enable the wealthy to show off their horses and apparel, Olmsted wrote in 1872 that his goal was to afford an outdoor experience to those of lesser means who had no opportunity to travel in the country.[2] Central Park launched a generation of city parks designed by the Olmsted firm, including Brooklyn's Prospect Park, San Francisco's Golden Gate Park, Philadelphia's Fairmount Park, Boston's Emerald Necklace, and the lakefront parks of Chicago.

In each of these park projects, Olmsted's biographer Elizabeth Stevenson explained, his purpose was not to preserve an existing landscape, but to create a restful and refreshing outdoor experience that would "make city life more civil, more healthful, more beautiful, more amenable to a good life for its inhabitants."[3] Thus the Olmsted parks movement, patrician in execution, was *humane* in objective. Indeed, Stevenson uses that very term to characterize his life and work: "Frederick Olmsted's ideal life was a humane, calm, free, and steady existence with time apportioned for quiet leisure as well as hard, engrossing work. This is the way he wished to live. By designing decent settings for work and play, he hoped to enable other people to live in this manner."[4]

With the onset of the City Beautiful crusade in the 1890s, Olmsted's humane populism was eclipsed by the Beaux-Arts-inspired enthusiasm for parks as civic ornaments, designed to inspire awe and cater to "those is easier circumstances." One of the signature parks of the City Beautiful era[5] was Chicago's Grant Park: a showplace of neoclassical formality designed by Edward Bennett that referenced not Central Park but Versailles and other European emblems of power and wealth. Let us now jump a century ahead from Grant Park to its new neighbor, Millennium Park, which might be considered the emblematic park of the Humane Decades.

Millennium Park: Way Beyond City Beautiful

The Art Institute of Chicago stands in solitary grandeur on Michigan Avenue, flanked by two parks that symbolize respectively the beginning and the end of twentieth-century urban design ideology. To the south of the museum lies the expansive Grant Park, the city's traditional "front yard" on Lake Michigan—an icon of City Beautiful aesthetics and paternalism. To the north of the museum lies compact Millennium Park, possibly the most exciting new urban park in the world—described by urban parks guru Peter Harnik as "Chicago's eye-popping showpiece playground between the Loop and Lake Michigan."[6]

I recently visited both of those parks on a hot July afternoon and experienced first hand the "sea change" described in the preface to this book, from the "Old Order" focus on how a park (or a city) *looks*—especially to persons with the right taste and breeding—to the "New Order" guided by how a park (or a city) *feels* to the general populace as it serves diverse needs. What a dramatic contrast unexpectedly struck me on that summertime stroll.

Both parks and the Art Institute itself are situated on land originally created when debris from the Chicago Fire of 1871 was dumped into nearby Lake Michigan.

As did Boston's Back Bay in the mid-nineteenth century and New York's Battery Park City in the 1980s, this windfall of new land presented Chicago with a unique opportunity to shape the future of its downtown. One of the first claimants to the new landfill was the Illinois Central Railroad (IC), whose mainline tracks since the 1850s have skirted the edge the downtown lakefront.[7] To this day the "train they call 'The City of New Orleans'" (memorialized by Arlo Guthrie) enters and leaves downtown Chicago—as do myriad commuter trains—via the old IC right-of-way extending south from the Randolph Street station on the post-fire landfill.

But apart from the railroad which was already there, wealthy Chicagoans whose mansions faced the lake opposed further development of the new land, citing an 1836 declaration by a state commission that the lakefront east of Michigan Avenue between Madison and 12th streets should remain "forever open, clear, and free." That doctrine was invoked at the end of the century in a lawsuit by the mail-order magnate A. Montgomery Ward, who successfully defended the view of the lake from his Michigan Avenue property—and the open character of the downtown lakefront—thus laying the legal basis for the future Grant Park.[8]

Consistent with Ward's wishes, the 1909 *Plan of Chicago* proposed creating a major lakefront park at the geographic vortex of the city where a park was already under development. Twenty years later, Grant Park opened to the public as the city's formal "front yard." As designed by Edward Bennett, the park was vintage City Beautiful: symmetrical, ornate, manicured squares of lawn and gardens, traversed by gravel walks and roadways, sprinkled with statues and monuments. At the geometric center of the park was Buckingham Fountain, a huge confection of marble nymphs and bronze dolphins set in a reflecting pool. The fountain's episodic geyser, illuminated by colored floodlights at night, has awed eight decades of Chicagoans and tourists. Elsewhere, the 375-acre expanse of Grant Park has accommodated athletic fields and outdoor festivals and concerts. Free evening concerts in the park with the skyline of Michigan Avenue in the background were favorite outings of my graduate school summers in Chicago.

However, on my 2011 summertime walk, Grant Park presented a very different and less inviting face. The immaculate flower gardens and lawns were still there, but they were cordoned off with dense hedges, fences, and plastic barriers (apparently left over from a recent festival). Buckingham Fountain itself was posted as off-limits to anyone wanting to cool off (fig. 8.1). More disturbing, there were hardly any people around. I bought a cold drink at a lonely concession stand and looked around for a shady place to sit. There were none: all benches beneath trees had been removed; the only benches remaining were in the open sun with few takers. I located a break in a hedge and stubbornly plunked down on a forbidden patch of shaded lawn. Moments later, a homeless man dragging a pink suitcase did the same. I asked him if he knew why there were no benches. His response, in essence: to discourage people like him from using the park.

Fig. 8.1. Chicago's Grant Park—a City Beautiful paragon—draws few visitors on a hot summer afternoon, in contrast with Millennium Park (figs. 8.2, 8.3).

Janice Metzger, in her intriguing book *What Would Jane Say?*, speculates about how Jane Addams and her settlement house colleagues might have responded to the *Plan of Chicago* in 1909 if they had been consulted. In relation to city parks, she imagines one of Jane's friends, Ada McKinley, commenting:

> I found the chapter on parks to be intellectually dishonest. [Burnham's] title is "The Chicago Park System." . . . Yet his emphasis in planning for the future is only on an upper-class playground on the lakefront and the suburban park system . . . [disregarding] over a million Chicagoans living in the great mid-section between the lakefront and the forests. . . . I'm not sure we should even participate in this exercise. It sounds to me like another of the business community's attempts to grab resources for their own use, and leave the rest of us to pay the bill.[9]

Metzger imagines that the settlement house women would have called for ample provision of neighborhood parks and playgrounds scattered around the city to serve the needs of ordinary families, especially immigrants and the poor who could not afford the cost of a roundtrip tram ride to visit a showplace park downtown.[10] I felt Jane Addams tap me on the shoulder that afternoon to underscore what the homeless man with the suitcase had said.

Millennium Park, a few hundred yards to the north beyond the great lion sculptures flanking the Art Institute's main entrance, is tiny compared with Grant Park—

a mere twenty-five acres on a platform above a parking garage ("the world's largest green roof" according to Mayor Richard M. Daley).[11] But this afternoon, unlike Grant Park, it was filled with kids and grownups strolling, sitting, wading, and people watching, all the while mesmerized by the amazing "Cloud Gate" and other park wonders. Although it is only 3 percent of the size of New York's Central Park, Millennium Park shares with Central Park an emphatic commitment to the welfare and enjoyment of the entire city populace—not just the elite clientele favored by City Beautiful park designs. Peter Harnik views Millennium Park comparable in importance to Olmsted and Vaux's masterpiece for revolutionizing the meaning and function of a city park: "Millennium Park has exploded onto the American urban park scene with an impact not felt since Central Park was unveiled in 1873. Numerous great parks have opened in the interim, but the Millennium phenomenon is due to a 'perfect storm' of location, artistic luminosity, politics and controversy, all greased by fantastic sums of money and fanned by the extraordinary publicity machine of America's most competitive city."[12] Harnik adds that Millennium "did more for the city's image (and probably for its real estate and tourism market) than any other development of the early twenty-first century—not to mention the ripple effect it had in other cities around the country."[13]

Millennium Park harks back to Robert Moses's early vision of decking over the railroad tracks along the Hudson River in Manhattan to create Riverside Park. The site of Chicago's new park was also previously a grim expanse of railroad facilities and parking. Harry Wiland and Dale Bell, recounting the genesis of Millennium Park in their book and PBS series *Edens Lost and Found: How Ordinary Citizens Are Restoring Our Great American Cities,* bluntly call the site at that time "one of the city's biggest eyesores . . . smack in the middle of downtown."[14]

Although Wiland and Bell's subtitle credits "ordinary citizens" for the impetus behind our new urban Edens, Millennium Park (like New York's High Line, discussed below) emerged principally from a coalition of strong city political leadership as well as generous corporate and individual financial support. A new 4,500-car garage built below grade level at Mayor Daley's instigation was decked over, providing an outdoor tabula rasa and a stream of parking revenue to help transform it into a park.[15] John Bryan, the retiring CEO of Sara Lee, served as the mayor's liaison to the business community, whose members poured hundreds of millions of dollars into the design and realization of the park. The architect Frank Gehry served as general adviser and designer of the Jay Pritzker outdoor performance pavilion and the BP Pedestrian Bridge connecting Millennium Park with the future Maggie Daley Park just to the east. The pavilion, with its extraordinary stainless steel lattice of speakers and lighting overarching a vast lawn, is the principal venue for the largest free concert series in the world hosted by Millennium Park.[16] The Nichols Bridgeway, leading to the Art Institute, literally connects the park to the cultural heart of the city. Access to the park by nonautomotive transportation is encouraged by the park's proximity to a commuter rail station and a bicycle storage facility equipped with lockers and showers.

The park's architecture is eclectic rather than uniform, ranging from nostalgic touches of City Beautiful to the ultra-modern feel of gleaming stainless steel and the latest in video-imaging technology. Cloud Gate—the park's most spectacular feature (fig. 8.2)—is a shimmering mirror-surfaced sculpture that reflects the city's skyline and people in its vicinity with both amazing clarity and distortion. Designed by Anish Kapoor and fabricated under the direction of Hollywood set designer Ethan Sylva, Cloud Gate (aka "the Bean" for its shape) is now a world-famous icon of the park and the city, designed to last for a thousand years.[17] Of equal popularity and technical virtuosity, the two fifty-foot glass block towers of Crown Fountain (fig. 8.3) bookend a reflecting pool from which splashing waders (or skaters in winter) are treated to ever-changing video images displayed on each tower—the faces of hundreds of ordinary Chicagoans spouting occasional cascades of water to enhance the fun in warm weather. Other park features include the 1,500-seat Harris Theater for Music and Dance, Lurie Garden designed by the Dutch landscape architect Piet Diblik (who also designed plantings on New York's High Line), and the McDonald's Cycle Center.[18] Within six months of opening in 2004, more than five million people visited the park, reveling in the variety of delights it offers.[19] At the park one freezing December evening, I watched over a hundred skaters circle the frozen wading pond, swaying to the strains of a Strauss waltz; like most of the park's attractions, skating is free of charge.

Millennium breaks the traditional park mold in many respects:

- The grass is meant to be sat upon (except when cordoned off to allow it to regenerate).
- A reflecting pond is meant to be splashed in or skated on.
- Lurie Garden, the park's primary botanical space, is planted with native prairie species and perennials to minimize maintenance and celebrate biodiversity of the upper Midwest.
- People are welcome to stay until late evening.
- Police are inconspicuous—security is provided by young people on Segways who smile and answer questions.
- Most signs are concerned with park features and special events rather than rules and regulations.
- Access is wide open with no gates or walls separating the park from the city.

Millennium is clearly a park for people, not just civic prestige (although that also is a major by-product). Despite its corporate provenance, it radiates the new spirit of humane urbanism. Both "Janes" (Addams and Jacobs) would be pleased.

Fig. 8.2. Cloud Gate—Anish Kapoor's mirrored sculpture reflects the city skyline and visitors to Millennium Park.

Fig. 8.3. The immensely popular wading and skating basin at Crown Fountain, Millennium Park.

Sea Change on the Hudson: Westway to High Line

Jane Jacobs's role during the 1960s in derailing the Robert Moses plans for highways across Manhattan and through Washington Square (discussed in chapter 3) certainly qualified her place in the pantheon of humane urbanists. But no sooner were those projects shelved than "Westway" sprang up to consume the energies of Jacobs's disciples during the 1980s. The defeat of that technocrat-era project cleared the way for two new parks on Manhattan's Lower West Side: Hudson River Park and the High Line linear greenway park.

Although much of Manhattan's waterfront originated as filled land, it lacked any equivalent to the "forever open, clear, and free" designation that applied to Chicago's downtown lakefront. Although that goal was sorely tested by local politics according to the historian Lois Wille,[20] it provided or at least afforded a presumption that the public interest in open space should prevail over commercial or residential development along Chicago's downtown lakefront.

New York City's waterfront in fact evolved under the opposite presumption, namely that its "highest and best use" was for maritime-related development including docks, warehouses, railroads, and eventually highways. As early as 1686, legal ownership of the waterfront to the low-water line was transferred by the British Crown to the fledgling new city of New York, which in turn sold waterfront lots to individuals "with the proviso that the owner must build the street and wharf," thus beginning the walling off of the city from its waterfront.[21] Maritime activities dominated the waterfronts of Manhattan and Brooklyn until the demise of passenger ships and the migration of cargo activities to the container ports in New Jersey and elsewhere. By the 1960s, most of the Manhattan waterfront lay derelict and abandoned, awaiting a new future.

The evolution of a post-maritime, mixed-use Hudson River waterfront began with the development of Battery Park City (BPC) adjoining the financial district in Lower Manhattan. The BPC site originated as a ninety-two-acre tract of landfill created with material excavated from the World Trade Center construction site in the early 1960s. Governor Nelson Rockefeller, a leading promoter of the World Trade Center, called for a planned mixed-use development to be constructed on the new land. The Battery Park City Authority (BPCA) was created by the state in 1968 to oversee the project as a public-private joint venture. But for a decade, writes Philip Lopate in *Waterfront: A Walk Around Manhattan*, "the project remained nothing but a sandy white beach . . . stalled by complexities of planning, bureaucratic rivalries, and New York's fiscal crisis in the 1970s."[22]

In 1979, after a prolonged and contentious design process, the state approved a master plan incorporating the BPCA's design guidelines for the project's construction. Battery Park City finally emerged as a multi-billion-dollar, mixed-use planned development that includes the World Financial Center, an upscale retail mall, various commercial and residential buildings, and a series of new waterfront greenspaces. (The BPC complex narrowly escaped destruction in the 9/11 attacks on the adjacent World Trade Center and was damaged by the hurricane storm surge of Sandy in 2012.)

Battery Park City is separated from the "real" city by an eight-lane traffic artery. Its World Financial Center is a consummate product of technocrat-era planning, which to Lopate "resembles nothing so much as an office park in Houston, . . . a superblock; four skyscrapers . . . turning their backs to the street . . . retail shops buried inside; and an overall visual monotony."[23] But BPC's waterfront esplanade and miniparks anticipate a more humane urban vision. After lengthy negotiation with civic interests, the BPC Master Plan required that at least 30 percent of the site would be retained as public open space. The nonprofit Battery Park City Parks Conservancy today operates a minipark system totaling thirty-six acres, including the riverside esplanade, the Robert F. Wagner, Jr. Park, gardens, walkways, and the 1.9-acre Teardrop Park, a meticulously designed green oasis completed in 2004.

Battery Park City soon spawned a much larger proposal for Westway—an interstate highway to be constructed on new landfill along the Hudson River waterfront with space for parks and real estate development on a platform above it. Federal highway trust funds were the proposed source of revenue for most of Westway; all levels of government and many civic organizations, including the prestigious Municipal Art Society, supported the project. Robert F. Wagner Jr. of the New York Planning Commission described the Westway enthusiastically in 1980: "Covering 4.5 miles of waterfront along the west side of Manhattan, the project would remove the abandoned piers in its path, add 182 acres of landfill, and remove the elevated structure of the West Side Highway, a major obstacle to waterfront access. . . . The [new] highway will be almost completely depressed and covered. Thirty-five acres of landfill will be available for residential construction, 97 acres for parkland, and 50 acres for commercial and industrial uses."[24]

But Westway was passionately opposed by a coalition of neighborhood and environmental interests led by the New York City Clean Air Campaign headed by Marci Benstock. Phillip Lopate recalls: "For once, Jane Jacobs and Robert Moses were in agreement. Both hated Westway."[25] After years of litigation, Westway ultimately was killed by a 1982 Federal District Court decree, which held that granting a landfill permit for Westway by the defendant "violated the National Environmental Policy Act, the Clean Water Act, and the Rivers and Harbors Appropriations Act." The court based its decision on the failure of the Westway sponsors to assess impacts of the project on striped bass habitat in the Hudson River Estuary, of which the area proposed to be filled they deemed a "biological wasteland."[26]

In a larger sense, Kent Barwick, president of the Municipal Art Society, characterized the outcome in a *New York Times* article as "a plebiscite on whether people prefer highways to mass transit"—over one billion dollars of federal transportation funding was reallocated from the highway to mass transit when Westway was defeated.[27] The historian Ann Buttenwieser pithily deems the outcome as "planning hubris met by community opposition."[28]

Community input helped to define a more limited proposal for Hudson River Park (HRP) that required no new landfill but proposed renovating some remaining abandoned piers for park and recreation space. With the backing of some Westway opponents, the state in 1998 created the Hudson River Park Trust as a state-city partnership to develop the park in collaboration with community groups. The HRP, under development at this writing, extends five miles from Battery Park City north to 59th Street. Much of its length is basically a double-lane paved bikeway threading between a highway and the waterfront. Its anchor feature is Pier 40, a massive enclosed structure providing both indoor and outdoor sports facilities. Other piers will be repaired and adapted to various recreational activities (including fishing), and a new 1,000-foot recreation pier (on a prior pier's footprint) is under construction at 44th Street. Once major capital projects funded by the city and state are completed, the park is intended to be self-funding from parking, office rentals, and other income sources. Reflecting the new awareness of the Hudson as a living estuary (not a "biological wasteland"), Hudson River Park is perhaps the first urban park to include an "estuarine sanctuary reserve" which occupies 400 acres of marine habitat bordering the shoreline.[29]

A few steps inland from Hudson River Park, a very different park is nearing completion: the High Line greenway, one of the city's most popular new outdoor amenities. As Hudson River Park reuses old piers, the High Line is recycling a former 1.45-mile elevated rail viaduct built in 1934 to remove freight trains from the city's streets. Rail use along the structure ceased in 1980 and the structure lay abandoned for two decades, slowly gathering a mantle of soil nurturing volunteer grasses and wildflowers. The highly photogenic juxtaposition of this "urban wilderness" with adjacent city buildings fired public interest in converting it into an elevated linear park.[30] It was also anticipated (correctly) that the project would enhance nearby real estate values. Friends of the High Line, a blue-ribbon nongovernmental organization, was formed in 1999 to promote this vision.[31]

Backed by $50 million in city funds and millions more in private funds, an international competition yielded a winning design by the architectural firm Diller Scofido+Renfro, selected from 720 entries representing thirty-eight countries. Within the confines of the narrow railroad structure, the High Line greenway (as largely completed in 2013), is a free public walkway that meanders among lush plantings of flowers, shrubs, and trees, providing close-up (second-story-level) views of the historic neighborhoods adjoining it, and breath-taking views of Manhattan and the Hudson River beyond (fig. 8.4).[32] As of May 2013, the High Line is open between Gansevoort Street to 30th Street. No visitor count is available, but the High Line, like Millennium Park in Chicago, is wildly popular as a free

Fig. 8.4. The High Line, New York City.

year-round outdoor park that celebrates people, place, and nature. Although it is largely dedicated to strolling (no bicycles or dog walking allowed), the Friends of the High Line hosts a smorgasbord of special activities which in 2011 included "a post-snowstorm Snow Sculpt-Off, a Salman Rushdie Karma Chain, rooftop dance performances, four competing teen step teams, mushroom-shaped bouncy houses, a temporary public plaza below the High Line, 15,000 roller skaters, avocado popsicles, a working water feature, kids releasing butterflies and earthworms, salsa dancing at sunset."[33]

The High Line is attracting huge investment in nearby condo development, retail shops, restaurants and wine bars, especially conspicuous in the newly trendy meatpacking district south of 18th Street. Chelsea Market, a vast retail and business complex in the former Nabisco bakery plant, adjoins the lower end of the High Line. Altogether, the High Line, like its counterpart in Chicago, synthesizes creative public-private financing, world-class design, and broad popular appeal. Together with Hudson River Park, it represents the ascendancy of people-oriented public space design over the earlier practice, represented by Westway, of viewing parks as add-ons to gain public support for highway projects.

Counterpoint: Boston's "Big Let-Down"

Boston's Big Dig involved exactly that strategy—using park space as bait to win approval of a highway project—but the project proved to be a bait and switch. The resulting Rose Fitzgerald Kennedy Greenway has been disappointing compared to such recent triumphs as Millennium Park and the High Line.

The saga of the Big Dig began with the construction of Boston's notorious Central Artery during the 1950s—an elevated highway slashing through the city's downtown. This highway, now a segment of Interstate 93, connects Boston's south and northwest suburbs, continuing on to New Hampshire. Like its counterparts in other cities, the project demolished a swathe of older commercial and residential buildings, effectively wiping out the city's West End neighborhood and displacing some nine hundred households. The elevated structure perceptually divided the historic North End and waterfront from the rest of the city, and the number of cars, trucks, and buses using I-93 every weekday vastly exceeded its planned daily capacity of 75,000.

As early as the 1960s, Boston's transportation commissioner, Fred ("Father of the Big Dig") Salvucci, began to investigate the concept of replacing the central artery with a larger-capacity highway tunnel. Over three decades, this concept was refined and promoted by Salvucci and his allies, gaining the support of several state governors. In addition to the burial of I-93, the Central Artery / Tunnel Project (dubbed the Big Dig by its promoters), would include a much-needed connection between the Massachusetts Turnpike and Logan Airport (the Ted Williams Tunnel) and a dazzling new bridge across the Charles River (the ten-lane, cable-stayed Leonard P. Zakim Bunker Hill Memorial Bridge). Perhaps mindful of the demise of Westway at the hands of public transportation advocates in New York, the Big Dig planners included a new rapid transit element—the Silver Line. And to win popular support, the entire project was to be draped with a twenty-seven-acre "greenway" to cover new surface area above the I-93 tunnel.

Anticipating a very costly project, the state agencies in charge of the Big Dig waged a lavish promotional campaign to sell the project to skeptical voters (who had defeated the I-95 Inner Belt in the 1960s, discussed in chapter 3). For instance, a glossy mid-1990s brochure distributed widely by a state/private partnership to promote the project gushed:

> The Central Artery / Tunnel Project will make Greater Boston even greater for the future. People will find the commute to their jobs faster and easier than ever; the city itself cleaner and greener, with significantly improved air quality. New England businesses will be able to ship products far more easily over our roads and through our airports and sea terminals....
>
> In fact, Boston will be a world-class city in every respect. The project will create more than 150 acres of open space and parkland, making Boston even more inviting to residents and tourists by providing easy access from the Downtown area to the Harbor.... Most important, the Central Artery / Tunnel Project will improve our quality of life by creating a spectacular city for our children and for generations to come.[34]

Leaving aside the dubious assertion about faster commutes (and whether in the 1990s commuters should be encouraged to drive to downtown jobs anyway), the outstanding impression of this and similar promotional material is how the promise of greenspace was expected to shield the rest of the project from public challenge.

But in contrast to the precise and complex engineering plans for the tunnel, bridge, and other hard infrastructure, the green elements of the Big Dig were remarkably vague. To be fair, the details of each park space were left for debate in hundreds of community meetings as the rest of the Big Dig took form. But as of 1998, a supportive essay in the *Boston Globe Magazine* imparted the impression that the project's greenspaces could achieve world-class importance, offering an array of activities on twenty-seven acres above the sunken artery, and thus become to Boston what the Mall is to Washington, DC, or the Champs-Élysées to Paris.

Oddly, much of the anticipated activity on the new greenway was projected to occur *indoors* within a winter garden proposed by the Massachusetts Horticultural Society, a history museum, and a new YMCA. Just how much greenspace would be left if those and other intended buildings were located there was never specified. The issue became moot, however, as Big Dig cost overruns (the project cost was estimated to be \$22 billion in 2008)[35] and the general economic downturn of the decade caused plans for those buildings to be abandoned.

Indeed, the entire Rose Fitzgerald Kennedy Greenway—as the proposed twenty-seven-acre surface was grandly named in 1995—is somewhat of an orphan, despite a 2004 promise from project managers to live up to their commitments for both dollars and greenspace acres, and to honor a 1991 commitment to retain 75 percent of that new space in nondeveloped status (i.e., open public park space). The Rose Kennedy Greenway Conservancy was established to plan and manage the greenway, wisely transferring that responsibility from highway engineers to professional park planners.

But the reality is a far cry from all the promises. The various park spaces are indeed taking shape, some with input from the adjacent communities. But the centerpiece Rose Kennedy Greenway has had a rough time getting "off the ground." A July 2008 lead story by Noah Bierman in the *Globe* (headlined "No-So-Green Acres") raised an early warning that the greenway might be headed for failure. Admitting that it takes time for trees to mature and other landscape features to become fully realized, the article nevertheless characterized the greenway as resembling "a fancy median strip—half its surface is paved."[36] Bierman quoted the Harvard landscape architect Michael Van Valkenburgh: "Generally, it's pretty ordinary and not what the city was promised 20 years ago." Furthermore, Van Valkenburgh commented, its design was "hamstrung by a lack of money and wound up with a 'meager park' that lacks shade and drawing power."[37] Contrary to the green swathe portrayed in Big Dig promotional materials, the greenway is bordered by multi-lane surface arteries and ruptured by cross-streets and ramps leading to and from the tunnel beneath it. Earlier plans to deck over the ramps for the buildings mentioned above have been shelved as too expensive, leaving the ramps and streets a

fertile source of traffic noise, pollution, and visual blight—exactly what the entire
Big Dig was intended to eradicate.[38]

In my personal judgment, in terms of both social and ecological function, and
based on a few strolls along the length of the greenway in decent weather, the park
can't match the grand slams hit by Millennium Park and the High Line. The Rose
Kennedy Greenway may be equivalent to Millennium Park in acreage and to the
High Line in length, but it lacks the creative, people-centered magnetism of those
parks. Some features of the greenway, like the hardscape plaza depicted in fig. 8.5,
seem to have emerged self-consciously from a design studio charrette.

The Rose Kennedy Greenway Conservancy, the new management entity for the
project, is striving to generate excitement through new signage, special events, festivals, and mobile food vendors.[39] The plantings, though dormant in winter, will
gradually settle in over time. But much of the "greenway" is pavement, slippery
in wet weather and blistering hot in summer. Outside the conservancy's control,
the visitor is beset with traffic—on the adjacent streets, cross-streets, and tunnel ramps. Pedestrians, runners, or cyclists encounter an endless series of slow
crossing signals. I came away with the strong impression—that the greenway was
mainly a selling point for the highway project, not a high priority in itself for public and private funding. I also suspect, perhaps cynically, that the anti-highway fer-
vor which spurred the fight against the Inner Belt three decades earlier was lulled
into complacency by the aggressive—and successful—salesmanship of the Big Dig
partnership of government and the downtown business community.

Chapter 9

Reclaiming Urban Waterways: One Stream at a Time

The river, then, much more than a passive and picturesque artifact in the landscape, represents meanings and values that are at the core of how we understand our place in the environment.

—EDDEE DANIEL, *Urban Wilderness*, 2008

Cities have long had a dysfunctional relationship with the rivers and streams that once nurtured them. Local waterways like Boston's Charles River or Houston's Buffalo Bayou historically linked ports with interior hinterlands, while also supplying inhabitants of their valleys with potable water, edible fish, water power, waste disposal, recreation and, in arid regions, irrigation. But by the late twentieth century, industrialization and urban sprawl had left countless urban streams polluted, channelized, dammed, diverted, and ecologically barren (see chapter 6). "Black bottoms" of physical and social malaise have festered in neighborhoods built along or above forgotten urban streams.[1] Even as federal and state clean water laws of the 1970s began to reduce point source pollution of major rivers and harbors, many smaller metropolitan streams, ponds, wetlands, and estuaries continue to be degraded by both point and nonpoint source pollution, remaining in noxious oblivion beneath the radar of federal and state authorities.

Stream restoration at the metropolitan scale eludes top-down, bureaucratic solutions due to the sheer complexity of balancing diverse geographic conditions, goals to be achieved, and multiple stakeholders—all of which vary widely from one stream and locale to others. Political fragmentation is another challenge: metropolitan streams cross political boundaries and often serve as geographic boundaries between municipalities, counties, or states. Also, metropolitan streams like the Milwaukee River or Washington, DC's Anacostia River typically flow through wealthy exurbs, middle-class suburbs, poor urban neighborhoods, and the central business district—thus physically linking disparate socioeconomic fragments of the metropolis.

In the absence of effective top-down intervention, the redemption of local water bodies has by default relied upon bottom-up grassroots efforts. Urban watersheds and streams, such as those discussed in this chapter, have been widely "adopted" by groups of local environmentalists, educators, recreationists, teachers, and others. Local activists who seek to restore urban streams for fishing, birding, canoeing, kayaking, and even swimming engage in activities such as trash collecting, money donating, lobbying, or demonstrating. These volunteer "troops" coalesce around river advocates and defenders who provide leadership, knowledge, persistence—and, sometimes, personal wealth and political connections—to harness local energy into effective restoration programs. Local initiatives have been stimulated partly by each other, especially within the same metropolitan area, and also by professional advocates in both the public and private nongovernmental sectors. Federal agencies which have contributed to local stream improvement projects include the National Park Service Rivers, Trails, and Conservation Assistance Program,[2] the Environmental Protection Agency Watersheds Program,[3] and the U.S. Geological Survey watershed database.[4] The Center for Watershed Protection founded by Thomas Schueler in Ellicott City, Maryland, since 1992 has been a source technical assistance and advocacy for urban stream restoration.[5] Schueler's West Coast counterpart is Ann L. Riley, a water engineer[6] and environmentalist with long experience as an advocate for urban creeks and streams in California.

Typically, local urban stream restoration efforts are triggered by an issue of urgent local concern such as flooding, pollution, fisheries depletion, ecological degradation, a proposed riverside development, or a need for aquatic recreation. Over time, the focus on the original driving issue may broaden into a more comprehensive watershed agenda. As illustrated in the four case studies that follow, the emergence of urban river stewardship reflects the very best instincts of grassroots humane urbanism and its focus on people, place, and nature.

Boston's Charles River: Inventing River Stewardship

Like the Boston Marathon, the Charles River extends from the western suburbs to downtown Boston, winding eighty river miles (versus 26.22 miles for the marathon) and dropping 352 feet in elevation. The 308-square-mile watershed, home to about 900,000 people, lies entirely in Massachusetts but is fragmented among thirty-five towns and cities (fig. 9.1). From its headwaters in the high-technology corridor along Boston's outer beltway (I-495), the river descends over a series of old milldams before entering a long stretch of gentle terrain and grassy wetlands occupying more than 10 percent of the watershed (about twenty thousand acres).[7] The downstream impoundment, known as the Charles River Basin, is a world-famous lagoon plied by sailboats and kayaks against the backdrops of Back Bay row houses on the Boston side and the campuses of Harvard and MIT on the Cambridge side. Since the 1930s, generations of Bostonians have celebrated the Fourth of July on the Charles River Esplanade to the strains of the *1812 Overture* supported by cannon and fireworks.

By the 1960s, however, the Charles was an ongoing environmental disaster—heavily polluted by industrial and sewage discharges, degraded as a habitat, repellant to recreation users, and a rising flood threat to low-lying areas. Flooding was the most immediate issue as suburban development covered the watershed with impervious land cover and structures impinged on floodplains. After a series of floods in the 1950s and 1960s, the region's political leaders summoned the Army Corps of Engineers to design a flood control project for the Charles. The Charles

Fig. 9.1. Map of the Charles River Watershed. Courtesy of the Charles River Watershed Association.

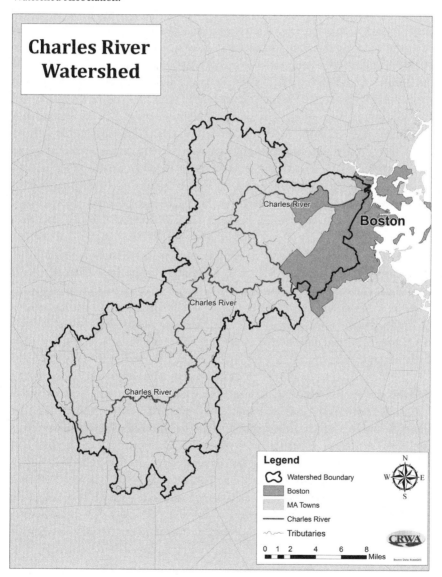

River Watershed Association (CRWA) was established in 1965 specifically to challenge the Army Corps's proposal to convert the still free-flowing river into a concrete flood channel. Rita Barron, a charismatic local resident and river activist, spearheaded the founding of the CRWA and its success in achieving a new vision of river stewardship, later serving as director of the organization from 1973 until 1988.

Fortuitously, the New England regional office of the Army Corps adjoined the Charles in Waltham, Massachusetts. Intense communication between the corps staff and CRWA produced a novel plan in 1970 to reduce flooding in the watershed through a three-pronged strategy: (1) acquisition and protection of several thousand acres of remaining wetlands for "natural valley storage";[8] (2) encouragement of land use regulation by watershed towns to limit further development in floodplain and wetlands; and (3) construction of a new dam at the river's mouth to alleviate overflow of the basin in Boston and Cambridge. While the third measure was unique to the geography of the Charles, the protection of wetlands and floodplains signaled a new model for application elsewhere. The Charles wetland acquisition project was authorized by Congress in 1973 and a decade later, the Army Corps had acquired about 8,100 acres of wetlands (in fee or easement), which were transferred to state and local authorities for management as natural flood storage and ecological restoration sites.[9] Concurrently, several watershed towns began to protect remaining wetlands from development through state wetlands laws.[10]

The protected wetlands serve as natural sponges to retain storm runoff that gradually seeps into river flow or into groundwater aquifers—alleviating flash flooding downstream that ensues from development and loss of wetlands. In addition, wetlands remove many sources of pollution by allowing microorganisms naturally present in the air, soil, and water to consume toxins. Because the wetlands were protected and allowed to perform their natural functions, the Charles River began to regenerate.

With the new strategy for flood hazard mitigation in progress in the 1980s, the CRWA focused on cleaning up the river. At the association's urging, tertiary wastewater treatment plants were constructed to reduce discharge of pollutants in the upper reaches of the river. In the mid-1990s, it persuaded the EPA regional director, John De Villiars, to notify some two hundred commercial dischargers that they were violating the Clean Water Act, threatening severe fines. This led to remedial action by the firms, further reducing pollution loadings.[11] CRWA also urged closure of landfills along the riverbanks. Combined sewer overflows (CSOs) were reduced and illegal hookups of sanitary sewers to storm drains were disconnected. As point sources were gradually improved, attention turned to nonpoint sources of pollution, primarily urban stormwater runoff. Clean-up efforts reduced runoff from roads and parking lots, improved septic system maintenance, restricted landscaping chemical usage, and curbed bank erosion through planting strips along riverbanks. CRWA seeks to arouse public awareness of the Charles River through education and outreach programs for community groups and schools. In cooperation with other river groups, CRWA coordinates Earth Day Charles River Cleanup. Its annual "Run of the Charles" canoe and kayak race attracts hundreds of boaters to the river every spring.

Ironically, water quality in the Charles has been adversely affected by efforts to improve water quality in Boston Harbor. In response to lawsuits charging violation of federal and state clean water laws, the state established the Massachusetts Water Resources Authority in 1985 to reduce the discharge of sewage and other wastes into the harbor. Diversion of storm runoff to the harbor via the new treatment plant reduces the normal flows in local streams like the Charles. During periods of hot weather and intense landscape irrigation, this loss of streamflow impairs stream water quality and biotic habitat.[12]

After four decades of advocacy and innovation, the Charles River Watershed Association is one of the leading organizations of its kind in the United States. When asked to name CRWA's proudest accomplishment, executive-director Bob Zimmerman responded: "Having the Charles be regarded as a treasure rather than a sewer."

Washington, DC's Anacostia: Waterfront versus Watershed

The Anacostia River has often been called Washington, DC's "other river" as compared with the much larger Potomac, which it meets within walking distance of the Washington Monument.[13] It is also called a sewer that has been a convenient receptacle for the wastes generated within its drainage areas. Until recently, the Anacostia was ignored and abused by the communities, industries, and public authorities that border it or share its watershed—Montgomery and Prince Georges counties in Maryland (upstream) and the District of Columbia (downstream) (fig. 9.2).

The look, feel, and condition of the Anacostia vary greatly depending on which reach is considered. The shorelines within the District of Columbia have been highly engineered with many floodwalls and storm outfalls. The Anacostia waterfront in Southwest Washington is partially occupied by the Washington Navy Yard, the Anacostia Freeway and other highways, the National Arboretum (except for visitors), a new baseball stadium for the Washington Nationals, and a general lack of parking and walkability. Upstream in Prince Georges and Montgomery counties, in Maryland, the river and its tributaries retain a somewhat more natural appearance.

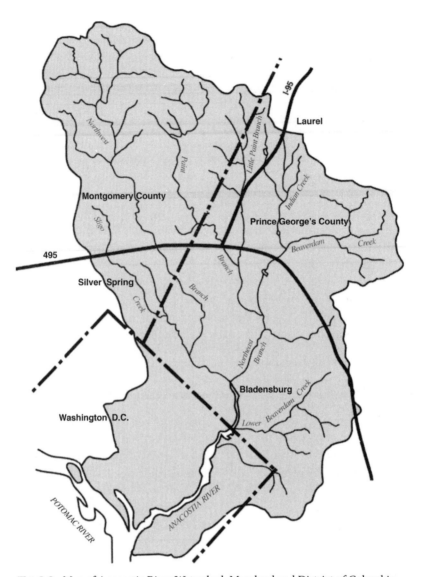

Fig. 9.2. Map of Anacostia River Watershed, Maryland and District of Columbia. Courtesy of the Anacostia River Watershed Society.

By the 1990s, the Anacostia symbolized the ignominious divide between the region's "haves" and "have-nots" in two respects. First, *environmentally,* the river receives sewage, sediments, chemicals, and other wastes from upstream Montgomery and Prince Georges counties and conveys them in ever-increasing concentrations through the impoverished neighborhoods downstream in the District of Columbia (along with sewage overflows, trash, and other waste discharged within the District itself). Second, the Anacostia represents a *socioeconomic* divide between its upstream and downstream communities, and also between the wealthy "western" and poor "eastern" neighborhoods of the District, the worst of the latter, as John Wennersten observes, being the Anacostia neighborhoods east of the river. "Life east of the river remained perilous in a community beset with not only a polluted river but social pollution in the form of chronic poverty, drugs, and violent crime. In April 2005, D.C. police reported that so far that year more than half the murders in Washington had been committed in Anacostia. As one young resident, David Smith, put it, 'Anacostia's always been a haven for the poorest people. This is where they dump their trash and dump the people, who I guess the city didn't want to see.'"[14]

For much of the twentieth century, the Anacostia was largely ignored. Those who lived within its watershed had neither the political nor the economic influence to affect any changes. Residents of more affluent regions on the western side of the city saw little incentive to take any action since they rarely ventured into Southeast Washington or Prince George's County. The years of extensive urbanization throughout the watershed left it highly degraded by the 1980s. Unmanaged development had resulted in habitat loss, erosion, sedimentation, wetland destruction, channelization, and toxic pollution. These impairments eliminated recreational opportunities such as swimming, fishing, and boating, and severed most ties between the river and its increasingly dilapidated communities and waterfront areas. While the Potomac River experienced a major rebirth as a recreational and an economic engine for the region, the Anacostia languished as an orphan on the impoverished eastern side of the city.

Over the past three decades, the environmental and social pathologies of the Anacostia have been increasingly publicized by DC-based national organizations like American Rivers and the Natural Resources Defense Council, and locally by the Anacostia Watershed Society (AWS), founded by Robert Boone in 1989. The AWS's primary efforts involve wetlands restoration projects, tree planting, water-quality monitoring and flagging, debris removal, community outreach, and watershed education. Nearly two decades later my colleagues and I wrote that "AWS provides some of the glue and a good bit of the vision for managing the larger watershed, working with many different groups and serving as an umbrella for smaller organizations."[15]

Most of the activities of AWS from cleanups to river tours involve an extensive network of volunteers, estimated to number over 41,000 people since the organization's founding. Thousands of youths have participated in the society's projects and programs. These engagements with children, Boone argues, are in fact "life shaping" and have the ability not only to change outlook on the river but also to fundamentally redirect young lives. AWS through its volunteers has planted of thousands of trees, stenciled more than 1,100 storm drains, and removed hundreds of tons of debris from the river. Its annual paddleboat regatta and weekly canoe tours have introduced thousands of people to the river and its varied environmental conditions and problems. Some 5,500 school children have participated in the Watershed Explorer's River Habitat education programs.[16]

AWS has engaged in environmental litigation using pro bono legal services from local law firms. It sued the District of Columbia Water and Sewer Authority (WASA) to expedite the construction of a vast project of stormwater detention and treatment known as the Long-Term Control Plan (LTCP). It has also sued EPA to hasten implementation of its Total Maximum Daily Load (TMDL) program for controlling nonsource pollutants in the Anacostia.

The Earth Conservation Corps (ECC), founded by Robert Nixon in 1992, is another key nonprofit working on the Anacostia. ECC recruits hundreds of youths from distressed neighborhoods as a hands-on workforce to reclaim the river, providing environmental education as well as leadership development and media arts training on site.[17] ECC over time has helped to monitor water quality, restore wetlands, tend a rooftop garden, and guide pontoon boat trip tours.[18]

As an inter-jurisdictional watershed, cleaning up the Anacostia has involved a quarter-century of efforts to craft partnerships among the various public stakeholders. In 1987, the District, the state of Maryland, Montgomery and Prince George's counties, and three federal agencies jointly established the Anacostia Watershed Restoration Committee (AWRC) to develop a watershed program to meet ecological, economic, and social goals for the river. By 1991, the AWRC had identified 207 stormwater retrofit, stream restoration, wetland creation, and riparian reforestation projects as elements of a proposed two-decade Action Plan for Restoring the Anacostia River.[19]

Concurrently, the District of Columbia under Mayor Anthony A. Williams (who served from 1999 to 2007) launched a new planning initiative embracing the Anacostia corridor within the city as a major redevelopment opportunity. A memorandum of understanding (MOU) was signed March 22, 2000, by eighteen federal and local agencies including the District, the National Park Service, the Office of Management and Budget, and the Army Corps of Engineers. The MOU established a "new partnership" that "envisions a new, energized waterfront for the next millennium that will unify diverse waterfront areas of the District of Columbia into a cohesive and attractive mixture of recreational, residential, and commercial uses by capitalizing on one of the City's greatest natural assets, its shoreline."[20] Notably, the Anacostia Watershed Society and other private interest groups were not part of the MOU and, not coincidentally, the MOU said little about cleaning up the river—a Herculean task that must be addressed at the watershed scale, not just within the District. The primary purpose of waterfront revitalization would appear to be real estate redevelopment.[21]

In 2003, the District of Columbia unveiled its Anacostia Waterfront Initiative (AWI). As featured in a major exhibit at the National Building Museum and accompanying media fanfare in early 2004, the AWI was received as a landmark in the city's history, comparable to the famous 1902 McMillan Commission Plan for the Washington, DC, Mall and park system. As noted, however, the initiative applies only to the District and does not address pollution loadings from the upstream suburban counties.

The AWI plan envisions a major geographic reorientation of the city: "With Washington's downtown nearly built out, the city's pattern of growth is moving eastward toward and across the Anacostia River. The destiny of the city as the nation's capital and a premier world city is inextricably linked to re-centering its growth along the Anacostia River and making its long-neglected parks, environment and infrastructure a national priority. The recovery of the Anacostia waterfront will help to reunite the capital economically, physically and socially."[22] These lofty goals would entail a thirty-year, $25 billion public-private investment to achieve five specific agendas: (1) a clean and active river; (2) eliminating barriers and gaining access; (3) a great urban river park system; (4) cultural destinations of distinct character; and (5) building strong waterfront neighborhoods.[23]

It is ironic indeed that the AWI should focus in part on the city's Southwest Waterfront—the locus of an early constitutionality challenge to the federal urban renewal program. As discussed in chapter 3, the U.S. Supreme Court in *Berman v. Parker* resoundingly approved the program: "It is within the power of the legislature to determine that the community should be beautiful as well as healthy, spacious as well as clean, well-balanced as well as carefully patrolled. . . . If those who govern the District of Columbia decide that the Nation's Capital should be beautiful as well as sanitary, there is nothing in the Fifth Amendment that stands in the way."[24]

Rarely does reality so poorly fulfill expectations. Far from "beautifying" the area, as David Moffat commented for the online forum group *Design Observer,* urban renewal of the 1950s and 1960s caused "large swathes of southwest Washington [to be] declared 'blighted,' and redeveloped with new mid- and low-rise housing [unaffordable to displaced residents] and commercial structures. About the same time, the area was also cut off from the nearby National Mall by construction of Interstate 395."[25] John Wennersten's pithy description concurs: "In its efforts to save the Southwest by bulldozing it, Washington urban redevelopment brought about the wholesale decline of the Anacostia community."[26]

The AWI Vision for the Southwest waterfront would renew the area once again, promising this time to create a "true urban waterfront where commercial, cultural, residential and neighborhood life can come together."[27] New public spaces would include a market square and a civic park as well as other smaller plazas and parks. A "grand public pier" would extend from the civic park. The existing waterfront promenade would be expanded and improved. New development would occur with six- to twelve-story buildings that would mix commercial and office uses with eight hundred units of housing. A new light rail line along Maine Avenue would serve and connect the Southwest waterfront. The plan also imagines water taxis and ferries.

The planning and management of the Initiative represents a tremendous political challenge in terms of reconciling the interests of AWI's eighteen member agencies, along with numerous other stakeholder groups, governmental bodies, and community groups. A specially created Anacostia Waterfront Corporation (AWC) modeled on the Battery Park Development Corporation in New York City was intended to spearhead the initiative. District government established the AWC in July 2004, but due to delays and political factors abolished it in 2007 with the AWI project assigned to the office of the Deputy Mayor for Planning and Economic Development.[28]

Understandably, a major concern among lower income residents of the Anacostia neighborhoods is the fear of gentrification, replacement of low-rent housing with high-end condos and apartments.[29] To address that concern, the AWI includes among major projects planned a new HOPE VI project to replace seven hundred existing units of public housing with a thousand new affordable housing units. It will incorporate some low impact development requirements and will facilitate connections to the river for residents of the project.

Another major issue for the District is, again, the combined sewer overflow problem. The city recently received $85 million from EPA to undertake initial planning for a new stormwater storage facility that many believe represents the best approach to addressing this problem; this facility would probably cost $1.3 billion, and would essentially constitute a large underground tunnel to store excess stormwater until it can be treated. This is seen as a more feasible alternative than investments to build a separated system that would likely cost substantially more (probably on the order of $4 billion). The city generally suffers from depending on an infrastructure seventy-five to a hundred years old that lacks major improvements made over the years.

The National Park Service owns considerable land along the east side of the tidal river in the District, as well as the extensive Kenilworth Gardens, Kingman Lake, and Kingman Island in the upper reaches. The potential impact of land use and management of these areas is great. The National Park Service is presently preparing its general management plan for the Anacostia to replace the existing one prepared more than twenty years ago. The NPS has also been actively involved in the District's Anacostia Waterfront Initiative (AWI) and anticipating projects that will contribute to or dovetail with AWI's plans. These include, for instance, an extensive trail system through the park that will dovetail with AWI's plans and connect with existing (or planned) trails in Maryland.

It is too early to assess the success of AWI since it is at least a thirty-year under-taking. Also, web-based computer graphics tend to blur what is "on the ground" and what is still in the mind of the landscape planner. Nevertheless, a Ten-Year Progress Report released in 2010 cheerfully recites progress on many fronts, in-cluding these six: (1) ten miles completed of the proposed twenty-mile Anacostia Riverwalk; (2) ninety-five acres of wetlands planted; (3) a 36 percent reduction in combined sewer overflows to the Anacostia; (4) substantial bridge and highway construction near the new Nationals Baseball Park; (5) start of construction of the 5.5-acre Yards Waterfront Park and mixed-use development (next to the sta-dium); and (6) substantial completion of the Marvin Gaye Park and Greenway in the low-income neighborhood east of the Anacostia. (See further discussion in chapter 10).

The Anacostia Watershed faces a daunting series of challenges to those who envision it as a cleaner, more accessible, and more ecological resource for metro-politan Washington, DC. One challenge, common to most urban watersheds, is its political fragmentation, namely among two Maryland counties and the District of Columbia, as well as among the many federal agencies that play various roles in the watershed. A second challenge is the multiplicity of problems, including poor water quality, loss of wetlands and other habitat, inaccessibility to local communi-ties, and the socioeconomic contrasts between suburban communities upstream and the District downstream. A third challenge involved the disparity between the District's Anacostia Waterfront Initiative and the broader aspirations for the entire watershed. Strikingly, the AWI is largely a governmental enterprise, based on the above-mentioned Memorandum of Understanding signed in 2000 by eigh-teen federal, regional, and District agencies with no private-sector participants. Efforts to rehabilitate the larger Anacostia drainage system, however, depend strongly on private-sector advocacy, especially by the Anacostia Watershed Soci-ety. And finally, there is the challenge of gentrification: if the District of Columbia Anacostia Waterfront Initiative is substantially realized, how will it affect the pres-ent residents of the lower income neighborhoods that now border the river?

Houston's Buffalo Bayou: From Concrete to Bioengineering

Houston, Texas, the nation's fourth-largest city, owes its location to the sluggish but occasionally wrathful Buffalo Bayou. Anglo settlement of the region originated at the confluence of Buffalo Bayou with White Oak Bayou around 1822. The city's founders, the Allen brothers, in 1838 laid out a grid street pattern oriented to the bayou and began to promote Houston as an alternative to Galveston as a port link-ing a vast interior hinterland of oak forests and grasslands with the Gulf of Mexico. In time, Houston thrived when cotton from the large plantations to the west was shipped from Houston and cattle drives from the prairie filled the stockades of the slaughterhouses. The destruction of Galveston in the 1900 Hurricane, and the discovery of oil at Spindletop near Beaumont, Texas, four months later ensured Houston's future as a major trading and transportation center. Throughout the 1920s and 1930s, oil refineries sprang up along the channel banks while other businesses directly tied to the petroleum industry began operation nearby. Today, the fifty-two-mile Port of Houston is the nation's leading port in foreign cargo and one of the largest ports in the world. Houston's population in 2010 was just over two million, while its metropolitan area population numbered 5.9 million, a 60 percent increase over 1990.

Sadly abused in the foregoing process was Buffalo Bayou itself which nurtured the early growth of the region. The bayou (or "creek" to non-Texans) meanders about seventy-five miles from its headwaters in the plains west of the city, through suburban Harris County and downtown Houston, to its mouth at the San Jacinto River, a few miles above Galveston Bay. Four-fifths of the 482-square-mile Buffalo Bayou / White Oak Bayou watershed is urbanized. Industrial wastes, combined sewer overflows, and nonpoint sources have long polluted Buffalo Bayou and its tributaries. High levels of fecal coliform in the watershed are yielded by storm water runoff from cattle grazing and wildlife areas, agricultural chemicals, and leakage from sewers and failing septic systems. The industrial corridor of Buffalo Bayou downstream from the city center is highly polluted with sediments containing toxic levels of dioxin, arsenic, zinc, and other chemicals.[30]

But the issue that focused public attention on the sad state of Houston's bayous was flooding, a result of both inland storm runoff and coastal storm surge (the latter worsened by land subsidence due to withdrawal of groundwater). The Buffalo Bayou drainage system is prone to flash flooding because of its prevalently flat topography and poor soil permeability. About 25 percent of Harris County has been estimated to be subject to a one-hundred-year flood. The area at risk has continued to expand further with widespread paving, building, and sewering, while land subsidence due to groundwater pumping has increased vulnerability to coastal flooding.

In the technocrat-era tradition, response to the flood hazard in Houston until the 1980s was predominantly top-down and structural. The Army Corps of Engineers and the Harris County Flood Control District constructed two flood control dams and reservoirs upstream in the Buffalo Bayou watershed and subsequently channelized over six thousand miles of local streams and bayous throughout the Houston region. In the absence of land use zoning in the City of Houston, areas behind levees were pervasively developed, and the streams themselves were rendered ecologically barren and treacherous to recreationists.

In reaction to ongoing channelization of the region's streams, local environmental activists and civic leaders formed the Bayou Preservation Association (BPA) in 1966 to serve as watchdog and advocate of alternative flood strategies. As early as

1967, BPA enlisted George H. W. Bush, then a local congressman, in opposition to further Army Corps work along Buffalo Bayou. BPA also proselytized the Harris County Flood Control District to explore bioengineering and land acquisition in preference to exclusive use of concrete structures.

Meanwhile, in the 1980s civic attention was drawn to the lower Buffalo Bayou as Houston's most prominent and neglected waterway. The Buffalo Bayou Partnership (BBP) was established in 1986 as a coalition of civic, environmental, governmental, and business representatives to plan and facilitate a greenway and related improvements along a corridor along the lower bayou ten miles in length and one mile in width, including the riverfront through downtown Houston. The BBP has since raised and leveraged tens of millions of dollars from private and public sources to advance its goals.

In 2000, the BBP board of directors decided that the new millennium was the ideal time to update planning for Houston's historic waterway. On behalf of the city and Harris County, the Buffalo Bayou Partnership oversaw the preparation of its master plan for the ten-mile segment of the Bayou extending from the western suburbs, through the city's downtown, and along the degraded industrial corridor to the bayou's mouth. The master plan was prepared by a team of landscape architects, planners, and economists as well as transportation, civil, and environmental engineers led by the Boston-based Thompson Design Group. Public opinion was solicited in workshops, focus group meetings, and design sessions. The $1.2 million cost to prepare the plan was shared equally by the county, the city, and BBP (backed by the business community and foundations). Additionally, the county's flood control district contributed $400,000 to develop a new hydraulic model to test the effectiveness of alternative flood management strategies for the bayou and its tributaries. The resulting master plan, *Buffalo Bayou and Beyond,* was released in 2002.[31]

In contrast with earlier City Beautiful–era park visions—such as a 1913 proposal for a linear park system along the region's waterways—the 2002 *Buffalo Bayou and Beyond* master plan maintains that nature is not a decorative ornament, but an essential system: "Buffalo Bayou's restoration will build value into Houston's urban economy; it will build a better quality of life to sustain and attract residents, celebrating the landscapes, wetlands and waterways beautifully integrated throughout the city."[32] The plan integrates environmental restoration, flood management, parks and open space, urban development, and transportation-related improvements, all to be executed over twenty years.

The overarching vision for the Buffalo Bayou is to restore it as an ecologically functioning system, to its place at the core of Houston's identity and sense of place.[33] The master plan balances conservation with development goals. It embraces enhanced quality of life with improved economic well-being for the residents of the regional community. Public improvements are expected to serve multiple objectives, such as a flood management project, which also provides recreation and open space and enhances natural habitat. Key goals include: (1) revitalizing the waterway itself, Houston's downtown, and nearby neighborhoods; (2) improving pedestrian and visual access; (3) reducing flood hazards; and (4) promoting cooperation between the public and private sectors. The public sector is expected to provide public improvements, such as streets, utilities, and public transportation, which in turn should encourage private sector investment.

Like the Anacostia Waterfront Initiative, *Buffalo Bayou and Beyond* seeks to promote real estate development and redevelopment within the river corridor. But unlike the AWI, the Houston project gives first priority to the ecological and recreational revitalization of the Bayou corridor to attract people and investment to downtown. (The AWI began with construction of the Nationals baseball stadium on the Anacostia).

Among its key environmental features, the *Buffalo Bayou* master plan called for public and private efforts to:

- Create "green fingers" along tributary streams to detain, filter, and cleanse stormwater.
- Reduce erosion and sedimentation by stabilizing bayou embankments.
- Develop a 2,500-acre park system along Buffalo Bayou linking Memorial Park in the west, via twenty miles of recreational trails to new parkland to be created in the industrial sector east of the city center.
- Regrade Buffalo Bayou's banks to widen floodplain and create space to restore wetlands, ponds, oxbows, and meander splays once found in the Bayou corridor.
- Reconstruct a major highway intersection to free up twenty acres of new open space, within which, a large pond could be created.
- Redevelop of derelict thirty-three-acre sewage treatment facility into an environmental awareness center.
- Build a new boathouse and boat launch ramps to enhance recreational boating on the Bayou.
- Promote "low-impact redevelopment" within the corridor including on-site stormwater storage and aquifer recharge through micro-detention features, such as rooftop gardens, vegetated swales, permeable paving (where applicable), and cisterns.

Since its release in 2002, visible progress has been made in achieving some of these goals. For instance, the new $15 million, twenty-three-acre Sabine to Bagby Waterfront Park involved drastic "bio-re-engineering" of the bayou channel through removal of floodwalls and restoration of a semblance of natural flow at the very doorstep to the Central Business District (fig. 9.3). The greenway on both sides of the stream provides flood storage when needed, and otherwise offers new

Fig. 9.3. Replacing concrete with greenways on Buffalo Bayou, near downtown Houston, ca. 2008.

hike and bike trail segments that connect to parks and bikeways elsewhere in the city. The park is embellished with innovative lighting that changes with the phases of the moon, a pedestrian bridge, replacement of invasive plants with native species, canoe launch ramps, and civic artwork.[34]

Immediately upstream, a much larger stream restoration and park project was launched in late 2011 along the Shepherd to Sabine reach, a 2.3-mile, 158-acre site.[35] And along the blighted industrial reach downstream from the city's business district, the future Buffalo Bend Nature Park is under development. Consisting of a series of graded, water-treatment wetlands, this park will help to provide East End residents with opportunities for hiking, bird watching, environmental education, and a sense of wildness otherwise lacking in this uninviting industrial corridor.[36]

Like the Anacostia Waterfront Initiative, the *Buffalo Bayou and Beyond* master plan is not a watershed plan since it focuses primarily on a ten-mile downstream segment in Houston. But according to BBP president Anne Olson:

> Improving Bayou water quality and regional biodiversity requires a planning scale much larger than the ten-mile Buffalo Bayou corridor, e.g., the Buffalo Bayou Watershed and the Greater Galveston Bay Ecoregion. To restore the Bayou to an ecologically functioning

system, the Bayou must be linked through open space corridors to other ecosystems and areas of diverse habitat. . . . The Plan envisions a conceptual plan for a "regional ecopark," connecting Buffalo Bayou to the Brazos River ecosystem, the Katy Prairie ecosystem, the Lake Houston and San Jacinto River ecosystem, and the Clear lake Recreational Area.[37]

Rio de Los Angeles: The Impossible Dream?

The Los Angeles River is what most of the nation's urban rivers would be if the Army Corps of Engineers had never met the Charles River Watershed Association. That collaboration of expertise and activism made it possible to nip the armoring of the Charles in the bud and, following suit, to keep the Anacostia relatively unchannelized (although heavily polluted), and now to tear out the concrete lining of the Buffalo Bayou near downtown Houston. But the engineering vision of the river as a concrete flood channel was never more perfectly achieved than in the concrete labyrinth that used to be the Los Angeles River. As described by the *New Yorker* writer Tad Friend: "The river is corseted in cement—poured in the 1930s [and later] by the Army Corps of Engineers to stop it from flooding—and further hemmed in by twelve crisscrossing freeways. Fenced off against would-be swimmers, its concrete banks tattooed with graffiti, referred to be local officials as a mere 'flood control channel,' the river is dry much of the year, and what water trickles through in high summer is mostly effluent from three sewage-treatment plants [and] 'urban slobber': storm-drain runoff carrying cigarette butts, pesticides, dog waste, Big Mac wrappers, and the occasional Yugo."[38]

From its headwaters in the hills above the San Fernando Valley, the main stem of the Los Angeles River flows fifty-one miles through fourteen cities (thirty-two miles in the city of Los Angeles), skirting the edge of L.A.'s central business district, south in company with the Long Beach Freeway (I-710) to its mouth at San Pedro Bay near the port of Long Beach. Over its course, the river drops 795 feet in elevation, causing it to be very flashy and historically destructive during heavy storms. The L.A. River's drainage basin, including nine principal tributaries, receives the discarded stormwater, sewage effluent, and trash from nine million people and myriad businesses, factories, and institutions that occupy the 850-square-mile watershed.

The Los Angeles River, like the Charles, the Anacostia, and Buffalo Bayou, gave rise to the city that later engulfed it. The city originated in 1781 with the founding of the Spanish pueblo at the river's confluence with Arroyo Seco, near today's Dodger Stadium and Chinatown. The early settlement was carefully laid out according to the Spanish "Law of the Indies" with a central plaza adjoined by civic and religious buildings beyond which lay a checkerboard of home lots and fields irrigated by ditches that conveyed water from the river. Unfortunately, a flood washed away the growing pueblo in 1815.

Exactly a century later, after a catastrophic flood in 1914, the Los Angeles County Flood Control District was established. Over the next five decades it gradually encased the river from headwaters to mouth in a trapezoidal "corset" of concrete lining both sides and most of the bottom of the channel. In some places the bordering flood control levees rise twenty or more feet above grade, blocking off any sight, access, or awareness of river to millions of Los Angelenos. By the 1980s, the district had constructed some 470 miles of stream channelization, 14 dams and flood storage reservoirs, 114 debris dams, 30 pumping stations, and 1,700 miles of storm drains—all designed to speed floodwaters out to sea and open up the Los Angeles basin to real estate development. The geographer Jennifer Wolch described the district as "both symbol and symptom of the twentieth-century approach to urbanization: a landscape rooted in worries about the destructive power of the natural world and the need to control it via massive engineering works and technological feats."[39]

Among many technocrat-era legacies, the site of the city's birthplace where the L.A. River and Arroyo Seco meet today, as characterized by Jennifer Price in the *Los Angeles Forum for Architecture and Urban Design*, is "one of the ugliest, most devastated spots on the Los Angeles River . . . a testament to the long-standing erasure of community, nature, and history in Los Angeles."[40] The site is scheduled to eventually become "Confluence Park," about which Price admitted, "you have to more than squint to imagine . . . you need special glasses, almost—but this is arguably the most logical site for a major city park in L.A."[41]

That contrast between the horrific present and an idealized future is a microcosm of the quixotic but determined quest to reclaim at least parts of the Los Angeles River from its current use as the region's main drain. In addition to potential ecological, economic, and recreational benefits, the river is increasingly regarded as, once again, a necessary source of fresh water—a benefit that it provided during the city's early growth but was abandoned with the development of external water transfers from the Owens Valley and the Colorado River, both now tapped out. Today, water management in Southern California is turning to the retention and storage of storm runoff in underground aquifers for future use, rather than simply discharging it into the Pacific Ocean. The district and other authorities in the area now operate over thirty-five "spreading grounds" where flood flows are diverted into groundwater aquifers. Additionally, an elaborate system of pipelines and wells operated by the district seeks to prevent saltwater from intruding into the aquifers beneath the watershed.

The "Dreamer in Chief" for the L.A. River is the poet and artist Lewis MacAdams, who founded the Friends of the Los Angeles River (FoLAR) in 1986 and has since been a soft-spoken but persistent agitator for public and private action to rehabilitate the river. Many other regional activists including artist Joe Linton, the author and illustrator of the FoLAR's official guide, *Down by the Los Angeles River*,[42] have joined MacAdams's efforts. The Friends organize annual grand clean-ups along segments of the river and sponsor many activities to promote public awareness of the river and its heritage.

FoLAR was to prove extremely effective in stimulating the region's byzantine political structure to address the river. In 1990 FoLAR persuaded L.A. mayor Tom Bradley to form the Los Angeles River Task Force, which in turn called for development of a river master plan by the relevant public authorities. Accordingly, in 1991 the Los Angeles County Board of Supervisors directed the county's departments of Public Works, Parks and Recreation, and Regional Planning to collaborate with the flood control district and other agencies in developing the plan. The National Park Service's Rivers, Trails, and Conservation Assistance Program provided technical assistance, and an advisory committee representing federal, state, county, and municipal agencies as well as nine regional environmental organizations was formed in September 1992. The ensuing *Los Angeles River Master Plan*,[43] as adopted by the county in 1996, outlined a multi-objective program for the river while affirming the primacy of flood protection. The Los Angeles County Department of Public Works as lead agency for the master plan reports the achievement of more than $100 million in projects along the River including the development of bikeways, pocket parks, landscaping enhancements, Earth Day events, acquisition of two State parks, and other community and environmental projects (fig. 9.4).

Frustrated with the slow pace of county efforts, FoLAR and City Councilor Ed Reyes persuaded Mayor Antonio Villaraigosa in 2002 to launch a more detailed study of the thirty-two miles of the river within the city of Los Angeles. The resulting *Los Angeles River Revitalization Plan,* adopted by the city in May 2007, set "intertwined goals: ecological restoration and management, the creation of recreational open space, economic development (where appropriate) and the encouragement of a sense of community around the river."[44] The 2007 plan lists 240 potential project areas in the city's stretch of the river, twenty "areas for targeted focus," five "opportunity areas"—all together involving hundreds of billions of dollars to be invested by 2032.[45] To accomplish this ambitious agenda, the 2007 plan called for three new institutions to be established:

1. *The Los Angeles River Authority,* representing the city, county and the Army Corps of Engineers with overall responsibility for projects in the river corridor.

2. *The Los Angeles River Revitalization Corporation,* a nonprofit charged with "promoting responsible development, redevelopment, and revitalization of properties along the LA River corridor [including] parks, open space, mixed-use buildings, retail opportunities, housing, and business space."[46]

3. *The Los Angeles River Foundation,* to raise funds to support activities outlined in the 2007 plan.

Fig. 9.4. Quiet time in a new mini-park adjoining the Los Angeles River, 2008.

Of this troika, the only unit established at the time of writing is the Revitaliza-tion Corporation. A first priority of the corporation, in cooperation with the state and river activists, is a complex plan to convert portions of Taylor Yard, a former railroad switching facility, into state park use. Taylor Yard is the largest undevel-oped parcel on the river, which it borders for two miles. Two parcels have been purchased by the state for park use: the forty-acre site of Rio de Los Angeles State Park, now under development, and a thirty-two-acre tract for the Los Angeles State Historic Park (known as the Cornfields). A lawsuit seeks to block commer-cial development of adjoining land within Taylor Yard, which river advocates hope will eventually be entirely acquired for restoration of ecological and recreational functions as the "Central Park" of the L.A. River corridor.[47]

Meanwhile, the densely urban San Gabriel River valley, east of the Los Angeles River watershed, is the focus of another L.A.-regional initiative led by the alli-ance called Amigos de los Rios. Amigos, a nonprofit organization devoted to building vibrant communities by designing quality healthy urban environments, operates in collaboration with local governments, businesses and residents.[48] Its affiliates include key public agencies, local officials, and regional environmental organizations. Like the Urban Resources Initiative in New Haven (which I discuss in chapter 10), Amigos works closely with local neighborhoods to create or revitalize small parks and public spaces. At a larger scale, the signature project is its own "Emerald Necklace," a seventeen-mile loop of parks and greenways now under development connecting ten cities and nearly 500,000 residents along the Rio Honda and the San Gabriel River.[49] This string of parks will provide or enhance recreational spaces for communities suffering from obesity, asthma, and land-use overcrowding.

Chapter 10

Humane Urbanism at Ground Level

> *City dwellers are happier and the city richer, finer, and more harmoniously balanced when two important ingredients are in place: sustainability and community. . . . On these two pillars rests all else.*
>
> —HARRY WILAND and DALE BELL, *Edens Lost and Found*, 2006

> *Nature . . . has unusually potent power to heal broken human landscapes and to humanize and reinvigorate distressed cities and built environments.*
>
> —TIMOTHY BEATLEY, *Biophilic Cities*, 2011

In 2009, a hefty and lavishly illustrated new book—*Mannahatta: A Natural History of New York City*[1] by ecologist Eric W. Sanderson of the Wildlife Conservation Society—transfixed the chattering classes of the nation's largest city. After feature articles about the book appeared in the *New York Times, National Geographic,* and elsewhere, and a summer-long *Mannahatta* exhibit at the Museum of the City of New York, it became an instant sensation. Its message: contemporary Manhattan occupies an island that hosted incredible biodiversity, forests, wetlands, ponds, and beaches at the time of Henry Hudson's accidental visit in 1609.[2] That natural landscape has been largely excavated, filled, dredged, drained, and buried—under a mantle of streets and buildings and a maze of subways and the sinews of gas, electric, water, steam, and digital cable that support life in the modern city. In *Mannahatta,* Sanderson stimulates awareness of the island's ecological past and documents the tentative first steps toward recovering and reconnecting some of the elements of that natural heritage:

> It is a conceit of New York City—the concrete city, the steel metropolis, Batman's Gotham— to think it is a place outside of nature, a place where humanity has completely triumphed over the forces of the natural world, where a person can do and be anything without limit or consequence. . . . But inside of New York another way of thinking is emerging, a new set of ideas and beliefs that . . . instead imagines a future where humanity embraces, rather than disdains, our connection to the natural world."[3]

Mannahatta represents a very recent and important perception of urban communities—namely as *human artifacts embedded within, suffused with, and dependent upon ecological systems and processes.* This recognition of course is scarcely limited to New York—that just happens to be where urban ecology is being intensely studied and documented by such leading research and education centers as the American Museum of Natural History, the Brooklyn Botanic Garden, the

New York Botanical Garden, and the Wildlife Conservation Society. New York is a vast demonstration project in the replanting and nurturing of native flora in the urban environment. One of the leaders in that effort is the New York Restoration Project, founded in 1995 by Bette Midler, which is rehabilitating parks in Harlem and the Bronx, promoting urban gardens, and helping to realize Mayor Bloomberg's MillionTreeNYC tree-planting campaign. Elsewhere, important centers of urban ecology research and teaching include Chicago's Field Museum, the Missouri Botanical Garden in St. Louis, the Cleveland Museum of Natural History, the Pennsylvania Horticultural Society, the Franklin Park Conservatory in Columbus, Ohio, the Houston Arboretum, the Los Angeles County Museum of Natural History, and so on.

Unlike the built environment, no two humane urbanist projects are identical. Goals and methods are mixed and matched in unique combinations in response to problems perceived and resources available. In other words, humane urbanism is an "ecological" phenomenon that adapts and mutates from one setting to another, with different cities and programs developing their particular blends of goals, methods, organizations, leadership, and funding sources. This chapter does not aspire to provide a comprehensive guide to the multifarious types of "down to earth" activities now in progress in cities across the nation, and around the world. That would be akin to trying to describe a forest, a native prairie, or a coastal estuary—where do you begin or end?

Fortunately, the ground has been well tilled by other writers such as those I quote in the epigraphs for this chapter. The broad terrain has been sampled in recent multi-author works such as *Growing Greener Cities* (2008), edited by Eugenie Birch and Susan Wachter, *Nature in Fragments* (2005), edited by Elizabeth Johnson and Michael Klemens, and *The Humane Metropolis* (2006), edited by me. Since Anne Whiston Spirn published *The Granite Garden* in 1984, the term "urban ecology" is no longer an oxymoron. Scholarly and popular works on urban ecology include *Understanding Urban Ecosystems* (2003) edited by Alan Berkowitz, Charles Nilon, and Karen Hollweg, *Wild New York* by Margaret Mittelbach and Michael Crewdson, *Wild in the City* (2011), edited by Michael Houck and M. J. Cody, and *Urban Wildscapes* (2012), edited by Anna Jorgensen and Richard Keenan. The urban biodiversity alliances Chicago Wilderness and its Houston counterpart described below have helped to foster a new awareness of regional ecological resources. Richard Louv's influential *Last Child in the Woods* (2005) has helped to raise public awareness of "nature-deficit disorder" resulting from growing up without contact with nature in everyday life. The list goes on.

The future of urban civilization, according to these and other interpreters of urban nature, may literally lie in the ground beneath our feet. We have long regarded that ground as mere space to be bought and sold, excavated, filled, paved, contaminated, built over and otherwise manipulated to accommodate the paraphernalia of our "built environment." Similarly, the networks of rivers streams, lakes, wetlands, estuaries, and aquifers that obstruct or threaten the development process have been channelized, buried, polluted, diverted, and rendered ecologically barren. The ever-quotable Lewis Mumford warned in 1956 how the modern city tends "to loosen the bonds that connect [its] inhabitants with nature and to transform, eliminate, or replace its earth-bound aspects, covering the natural site with an artificial environment that enhances the dominance of man and encourages an illusion of complete independence from nature."[4]

We are finally beginning to overcome that illusion and accept our dependence on nature and ecological services, not just "out there" beyond the metropolitan fringe, but at our very doorsteps. This acceptance involves, in part, recognition of the quantitative values of ecological services, as documented by the ecological economist Robert Costanza and his colleagues. But whether or not reduced to dollars, the qualitative value of urban nature is also gaining influence, as in this statement by urban naturalist Michael W. Klemens:

> Biodiversity is inextricably part of our sense of place, the very fabric of our comfort and our "being" at a particular locus. The natural world provides the texture and variety that define where we live, work and play. So defined, biodiversity is the tapestry of colors on a wooded hillside in October, the interplay between water and reeds, the chirping of crickets on a summer's night, the ebb and flow of natural systems evolving over time. And it is that natural template, the very foundation upon which our society is built, that I define as biodiversity, or more simply stated, nature.[5]

While urban design professionals manipulate the physical form and appearance of the built environment (whether or not in accordance with green building principles), another form of adaptation to the enveloping metropolis is emerging, one which focuses on the unbuilt elements of the urban environment (sometimes laboriously called "green infrastructure.") Such adjustments are concerned less with the way urban places look and more with the way they work, ecologically and socially.

Fundamental to this new perspective are five premises: (1) metropolitan regions are essentially inescapable, so we might as well make them as habitable, safe, and pleasant as possible; (2) the preceding observation applies across the socioeconomic spectrum to rich and poor alike; (3) the laws of nature are not suspended within urban areas; (4) respecting and restoring natural systems within urban places is often more cost-effective than using technological methods; and (5) sharing "down to earth" activities like urban gardening, native plant restoration, stream clean-ups, and tree planting brings people together and builds a sense of community.[6]

In this final chapter, "Humane Urbanism at Ground Level," I do not offer a compendium of "why" and "how" to reconnect with nature in the urban environment; such a resource is better provided by the writers listed above and others. Rather, my intent here is to "ramble" without a fixed itinerary, road map, or GPS among diverse examples of ground-level humane urbanism in various localities. The goals and methods of different projects are less important than to recognize the diversity, energy, and flexibility of such initiatives. Numerical results in terms of, for example, trees planted or saved, wetland acres restored, vegetables grown, bike-way miles added, invasive species removed, fish stocks revived, songbirds counted, and bugs discovered by children are less important than the halo of good feeling and sense of community that comes from direct personal contact with nature and each other. This chapter will sample experience in various cities at scales ranging from a quest to save a single tree to regional alliances.

A Tree Grows in Poughkeepsie

Poughkeepsie, New York, calls itself "The Queen City of the Hudson," an echo of its heyday in the early twentieth century as the leading retail and business center of the Hudson Valley between New York City and Albany.[7] After World War II, the city lost population and wealth as urban renewal and highway construction ripped out its older neighborhoods and suburban sprawl in the *town* of Poughkeepsie (a separate legal jurisdiction that surrounds the *city* of Poughkeepsie on three sides) hijacked its white middle-class and retail patronage. Adding insult to injury, the computer giant IBM established its headquarters in the town in the 1940s where land was cheap and nonwhites invisible. The immense tax base represented by IBM and spin-off growth thus accrued to the *town,* not the *city* (which houses most of the region's nonwhite and low-income households). The city lost about one-quarter of its population between 1960 and 1990, dropping to 28,800 as the town soared to 40,100.[8] As I stated in chapter 4, Poughkeepsie is a glaring example of the ironclad rigidity of local political boundaries that became set sometime in the past for reasons long forgotten but endure as Chinese Walls between separate units of local governance, inexorably shaping the checkerboard of municipal winners and losers.

But this battered old river city of only 5.7 square miles with rows of empty downtown storefronts qualifies to open a chapter called "Humane Urbanism at Ground Level." Poughkeepsie is slowly clawing its way out of its malaise (as possibly attested by its population increase to 32,700 in the 2010 census). An early accomplishment of that tentative recovery has involved the saving of a single tree, the Forbus Butternut, whose imminent demise stimulated perhaps the smallest-scale victory in the annals for humane urbanism.[9]

The tree in question is situated in a residential neighborhood of Poughkeepsie surrounded by comfortable, postwar middle-class homes. The tree itself is about seventy-nine feet tall, with a trunk diameter of over four feet (fig. 10.1). The naturalist Rutherford Platt (my father) described attributes of the species:

Fig. 10.1. The rescued Forbus Butternut tree in Poughkeepsie, New York. Courtesy Harvey Flad.

> Butternut (*juglans cinerea*) . . . has the face of a solemn camel with a downy brow be-
> low each bud. . . . Butternut grows singly in rich hillside pastures and along roads in
> New England and central New York State. It's a small version of black walnut, with light,
> feathery leaves and black branches—an intimate tree compared to the giant black wal-
> nut. You can reach its lower branches to see the faces and hairy twigs, and smell its leaf
> fragrance. . . . Butternut kernels have the highest food value of any of our nuts.[10]

This particular Poughkeepsie butternut is a remnant of a grove of old trees adjoining a historic home that once stood on a nineteenth-century farm just outside
the city. Once annexed to Poughkeepsie in the early 1900s, portions of the property were gradually sold for home sites and for the city's high school complex.
By 2000, a remaining thirteen-acre parcel—upon which the home and adjoining
butternut stood—was approved for a residential subdivision (a sign perhaps of
Poughkeepsie's improving fortunes).

It is not unusual for homeowners to enjoy the benefits of nearby open space
and trees owned by someone else, but that does not obligate the latter to preserve
such amenities without compensation. Even significant trees like this butternut
(or homes listed on the Historic Register as in this case) are rarely protected by
local ordinance from being removed by their owners. That is exactly what was
about to happen on a warm, sunny day in July 2000 when a neighbor across the
street, Robin Poritzky, intervened—with a pitchfork. Regardless of legalities, wrote
Harvey Flad and Craig Dalton, the authors of a case study about the tree, it "had
become a *place* in the cultural landscape of the city."[11]

Legalities, however, usually trump landscape and culture—as with myriad historical buildings and great trees lost to bulldozers and chainsaws over time. That is
exactly why the subtitle of this book refers to "The Struggle for People, Place, and
Nature." Humane outcomes in the contentious urban environment do not simply
happen: they have to be won.

As Flad and Dalton tell the story, Robin Poritzky and her neighbors quickly
started to collect information about the tree from various sources. A huge boost
came from a state forester who declared this tree to be the "State Champion Butternut"—as well as a listed endangered species. Publicity of the tree's status brought let-
ters of support from regional environmental organizations like Scenic Hudson and
Pete Seeger's Hudson River Sloop Clearwater group. The neighbors, now gaining
confidence, incorporated the Forbus Butternut Association (FBA) in order to attract
financial support. The tree's owner helpfully refrained from cutting the tree down,
(unlike Bethlehem Steel, which bulldozed its huge tract of the Indiana Dunes to
stymie the "Save the Dunes" campaign in 1966, as discussed in chapter 5).

In 2002, the new mayor and FBA ally, Colette Lafuente, persuaded the city council to approve $10,000 toward the purchase of the site as a city park. With addi-
tional private funds and a state grant of $32,750, the association assembled $60,000
to purchase five of the six available lots. The outcome was the establishment of

Walkway over the Hudson State Park

Poughkeepsie offers other promising stories of humane urbanism in progress. One is the rehabilitation of Fall Kill Creek, which flows generally from Eleanor Roosevelt's former Hyde Park residence ("Val-Kill") to the Hudson.[1] Another homegrown project is literally bridging Poughkeepsie's past and future.

In 1889, the Poughkeepsie-Highland Railroad Bridge opened as the longest bridge in North America and the only Hudson River span south of Albany. It long served as a vital freight and passenger rail link between New England and the Midwest, as well as within the Hudson Valley, until 1974 when it closed after a fire.[2] In 1992, Walkway Over the Hudson was formed as a local nonprofit organization to stimulate the conversion of the 1.2-mile level bridge surface into a pedestrian and bike path high above one of the nation's iconic rivers.[3] In 2009—seventeen years later and at a cost of $38 million, with major funding and professional support from the Dyson Foundation in Millbrook, New York—the bridge reopened as the Walkway Over the Hudson State Historic Park. Now paved, lighted, and handicapped accessible, the span is one of the world's longest and highest walkways—at 212 feet above the river, and also crossing high above downtown Poughkeepsie with outdoor market stands at its starting point (fig. 10.2).[4] With planned connections to regional bikeway systems and a future twenty-one-story elevator to link the city's riverfront with the Walkway deck, this project is a major stimulus to the economic and cultural revival of Poughkeepsie. During its first year of operation, the Walkway hosted 780,000 visitors, far exceeding expectations of its backers. At this writing, the Dyson Foundation is negotiating with the city to purchase a 2.7-acre city park at the base of the elevator near the site of the city's original settlement to connect the Walkway and the reviving Poughkeepsie riverfront. (As of 2012, a similar trans-Hudson bike and walkway has been proposed for the obsolete Tappan Zee highway bridge across the Hudson just north of New York City.)

1. Information from www.clearwater.org/ea/ej/watershed-management/fallkill/.
2. Carleton Mabee, *Bridging the Hudson: The Poughkeepsie Railroad Bridge and Its Connecting Rail Lines: A Many-Faceted History* (Fleischmanns, NY: Purple Mountain Press, 2001).
3. Walkway over the Hudson, "General Information," available at http://walkway.org/visit-walkway.
4. Flad reports that Walkway visitors marvel at the view of the city and its riverfront, thus contributing to the rebound of Poughkeepsie's reputation. (Personal communication, March 6, 2012).

1.3-acre Forbus Butternut Park, a joint city-state park managed by the FBA—very likely the smallest park in the New York State system. The tree thrives today amid its patch of greenspace, a permanent contribution to the people, place, and nature of Poughkeepsie.

Replanting "Elm City"

New Haven, Connecticut—colonial port city, former manufacturing center, boyhood home of the theologian Jonathan Edwards, and home of Yale University—is

Fig. 10.2. A vendor's stand at the Poughkeepsie end of the "Walkway over the Hudson" (the author's wife, who loves kale, is at left).

known as "Elm City" for the rows of stately elms that once lined its downtown streets. Commemorated in the maudlin Yale Glee Club standard "'Neath the Elms," most of them had succumbed to hurricanes, street widening, urban renewal, and Dutch Elm Disease by the 1960s when I was a student there. As many New England cities and towns learned in the Great Hurricane of 1938, ornate street trees of the same species and age tend to disappear at the same time, leaving a great emptiness and loss of shade.

New Haven was briefly a poster child for the federal urban renewal program under the direction of Edward Logue and with the keen participation of Yale.[12] By the late 1960s, however, the city's downtown was a wasteland of highway ramps, brutalist-style office and sports facilities, banal apartment buildings, garages, and the inevitable in-town shopping mall (Chapel Square), which soon folded. From its peak population of 164,000 in 1950, the city dropped to 126,000 by 1980 with the departure of middle-class white families to surrounding communities. Currently the city's population of 129,000 is about 37 percent African American and 21 percent Hispanic. The city is the second poorest in Connecticut (behind Hartford) with a poverty rate of 26.7 percent in 2009.[13] Not surprisingly, the exodus of families before the 1990s left New Haven's older residential neighborhoods awash in shuttered homes and vacant lots.

Enter the Urban Resources Initiative (URI), a collaboration between the Yale School of Forestry and Environmental Studies (F&ES), the City of New Haven, and the Community Foundation of Greater New Haven. URI's original mission, which continues today, was to convert vacant lots, one at a time, from eyesores to neighborhood amenities through cooperative projects with local neighborhood groups. Under the URI model, Yale students work with local residents to discern how a particular vacant lot (city-owned due to tax foreclosure) might be redesigned for community use as a food or floral garden, a minipark and gathering space, or some combination thereof. The students then help design the site and work side-by-side with local residents to plant trees, lay out gardens and paths, and otherwise achieve the neighborhood's vision. Maintenance of each site thereafter is a community responsibility. Since 1995, URI has facilitated about two hundred urban restoration projects (fig. 10.3).[14]

Fig. 10.3. Tree planting under the auspices of Urban Resources Initiative in New Haven, Connecticut. Courtesy URI.

Environmental education is a second URI priority. Its Open Spaces as Learning Places program draws on the city's parks, ponds, cemeteries, and other greenspaces as resources for teaching children awareness of nature around them. In 2001, URI began efforts to develop a sixth-grade curriculum with both classroom and outdoor elements. Yale F&ES students taught the curriculum to dozens of classes until it was officially adopted by the New Haven Board of Education for the school district in 2009. A cadre of teachers and administrators trained by URI then assumed responsibility for the ongoing program.[15]

Since 2007, URI has assumed a major role in achieving New Haven's goal of restoring the city's depleted tree canopy by planting ten thousand new trees in public spaces and on private property. URI is providing technical assistance on the selection of trees by size and species in accordance with site constraints and the need to diversify the composition of the urban forest. URI's GreenSkills program trains high school students and ex-offenders (as separate crews) in skills of tree planting and maintenance. More than 2,000 trees have been planted by teams of trainees and Yale students in public spaces (i.e., parks or greenstrips between streets and sidewalks). At the same time, dozens of trainees have received marketable training in arborculture and landscaping skills.

The effectiveness of New Haven's URI program may be a result of the following four factors: (1) a strong ongoing partnership between the city, the university, the community foundation, and community groups; (2) stable leadership through its CEO, Colleen Murphy-Dunning, who has directed the program since its founding; (3) a focus on New Haven rather than on scattered efforts throughout a wider region; and (4) sharing experience with peer programs through an eight-city Urban Ecology Collaborative.

Nuestras Raices: Holyoke's Homegrown Revival

Just a few miles downstream from my city of Northampton, Massachusetts, the Connecticut River flows past Holyoke. Two more different urban neighbors could scarcely be imagined: Northampton, a small but bustling educational, retail, professional, and cultural hub, and Holyoke, a former industrial powerhouse reduced in some sections to the equivalent of "postwar Dresden."[16] The contrast is mostly due to a fluke in physical geography: Northampton is bordered by a broad level floodplain that is still devoted to farming even though the town center, removed from the river, has organically evolved with the times. Holyoke, by contrast is located smack along the river, which is confined by uplands and drops sixty feet in elevation—an ideal site for a waterpower dam. Holyoke Dam and its canal system harnessed the river in the mid-nineteenth century; the city, which arose to house mill workers and company executives, was one of the nation's most renowned planned industrial cities. But the prosperity that the city gained from its dam was later lost as the paper industry, the city's primary economic base, moved elsewhere. Like much of industrial New England, Holyoke was left with abandoned industrial buildings and tenements in "the flats" near the river, a small heritage of landmark civic and religious buildings, and assorted Edward Hopper–vintage mansions on the adjoining hillsides.

Fast forward to 2010, when Holyoke's population stood at about 40,000, a loss of 26 percent since 1950: now 41 percent of Holyoke residents are Latino, many of whose families migrated to the Pioneer Valley as agricultural workers from Puerto Rico over decades. About a quarter of Holyoke's population live below the poverty line. Holyoke's business district, which in the 1950s flourished with department stores, cinemas, and "five-and-tens," is pockmarked with vacant storefronts and offices. (At least Holyoke gains taxes and jobs from a regional mall within the city limits near Interstate 91.)

Agriculture, which originally brought many of the city's Latinos to Holyoke, has helped to spur the city's incipient revival. In 1991, a group of neighbors began farming a small plot of land they called La Finquita (Little Farm). According to Catherine Tumber, "their purpose was not only to grow food but to transmit the agricultural skills they had acquired in Puerto Rico to the younger generation."[17] This group, which called itself Nuestras Raíces (Our Roots), evolved under its longtime director Daniel Ross into a national model of grassroots self-help and humane urbanism (fig. 10.4).

The farming side of the Nuestras Raíces (NR) program now involves a small greenhouse, a thirty-acre farm (La Finca) and a network of family-operated community gardens scattered around the city. The greenhouse at NR's central office location is equipped with an aquaponic system and large platforms for starting vegetables from seeds. NR distributes seeds and young plants to all comers free of charge with advice on how to nurture them. The farm is a "business incubator" where city residents are trained in farming skills and, when certified, may be allotted a plot to manage within the farm. Many of these individuals have then moved on to start commercial farming and marketing enterprises on their own. The community gardens provide more space for growing food in local neighborhoods, with much produce aimed at a Puerto Rican clientele sold in seasonal farmer markets and local grocery stores. In addition to farming skills, NR offers workshops to local youths and adults on life skills, business practices, and health issues such as obesity and asthma management.[18]

Nuestras Raíces has also fostered spin-off businesses that provide jobs to young people from Holyoke and nearby communities. Its El Jardín Bakery began as a source of baked products for a restaurant that NR once operated at its downtown premises. The restaurant eventually closed but the bakery incorporated and moved thirty miles up the valley to South Deerfield, from where it now distributes high-end sourdough bakery products throughout the valley. Another very

Fig. 10.4. Growing food and community at Nuestras Raices, in Holyoke, Massachusetts. By permission of Nuestras Raices Inc.

promising spin-off is Energia, a for-profit company located in downtown Holyoke whose youthful employees conduct home energy audits, install insulation, seal leaky windows and doors, among other energy conservation services. By 2010, Energia was partnering with the Holyoke Housing Authority to install solar panels and insulation on many of the city's older public housing projects.[19]

Washington's Marvin Gaye Park

Watts Branch, a humble tributary to the Anacostia River in Washington, DC, is not exactly a mecca for visitors to the nation's capital.[20] This muddy little creek wanders from its origin in Prince George's County, Maryland, into the Anacostia neighborhood of Northeast Washington—one of the District's poorest neighbor-hoods, separated from the rest of the city by elevated highways, railroad tracks, a swathe of National Park Service land, and the Anacostia River. Many of the streets near the creek are lined with boarded-up structures, vacant lots, public housing, few businesses, and many gospel churches. An eight-story concrete high school building, half its windows boarded up, looms like a prison block above empty sports fields enclosed by a chain-link fence. The District's Seventh Ward, which includes the Anacostia neighborhood, is plagued by crime, drugs, and joblessness.

A highly polluting power plant, a huge brownfield, and an old city landfill near the Anacostia have further blighted the vicinity.

The Watts Branch area, however, has rich associations with Washington's African American culture and history. Suburban Gardens, a blacks-only amusement park before World War II, was located near the creek. Martin Luther King Jr. spoke in the community in 1961, an event now commemorated by the newly established King Nature Sanctuary. In 1966, Lady Bird Johnson recruited the landscape architect Ian McHarg and the philanthropist Laurance S. Rockefeller to help revive the linear park adjoining Watts Creek where she launched the urban component of the Johnson administration's Keep America Beautiful Initiative. That park, now renamed Marvin Gaye Park to honor a renowned musician with local roots, forms the "green spine" of neighborhood improvement efforts today. The Crystal Lounge where Gaye first performed has been converted into a community facility (called the Riverside Center), for classes and youth training. Nearby, Nannie Helen Burroughs Avenue, named for the founder of a female black Baptist school, has been designated for major upgrading under the District's Great Streets Program. HOPE VI and Habitat for Humanity are building new residential units, and public housing renovations are underway.

Along with the District government and local organizations and churches, the catalyst for much of this activity is Washington Parks & People (WPP), an NGO established in 1998 by Stephen W. Coleman, who serves as its executive director. Coleman's mantra is: "Reconnect people with the land and use the land to reconnect people with each other."[21] Under leadership from WPP, and partnerships with public agencies, local schools, and nonprofits, some five hundred trainees and many volunteers since 2001 have removed tens of thousands of bags of garbage, three thousand tires, fourteen thousand hypodermic needles, dozens of abandoned cars and trucks, and over six million pounds of bulk trash and debris. WPP and the District have jointly created a 1.5-mile paved walk and bikeway along the stream, and have overseen the planting of more than two thousand native trees and shrubs by young trainees, with many of the trees now grown in the program's own nursery (fig. 10.5).[22] Crime has been reduced, stream water quality has improved, and people are beginning to use the park spaces along Watts Branch rather than shunning them. WPP also helps to preserve the area's black cultural history by providing programs and educational signage, and through recording of oral history.

On a visit to Watts Branch in May 2011, I encountered a crew of teenagers engaged in planting sizeable trees with heavy root balls, creating an instant "urban forest" along the newly paved bikeway that follows the creek toward the Anacostia. I followed the walkway to the Riverside Center, where Marvin Gaye began his musical career—now converted by WPP into a colorful hive of classes, meetings, exhibits, and general commotion of busy young people. My host however, a WPP

Fig. 10.5. Tree nursery at Marvin Gaye Park, Washington, DC.

staff member, admitted that what the Watts Branch neighborhood desperately needs, in addition to better schools, housing, and jobs, is a decent grocery store—th̶ ̶ ̶ ̶ ̶orhood has long been a "food desert." There is also worry that the Dis-̶ ̶ ̶ ̶.̶acostia Waterfront Initiative (see chapter 9), including the Watts Branch ̶ighborhood, may attract new upscale development and gentrification to the community, forcing low-income residents of the area to yield once again to power and money.

Two Acres in Milwaukee: The Growth of "Growing Power"

One bright spring morning in 2005, I join a group of fifth graders and their teachers for a tour of a two-acre agricultural complex on the west side of Milwaukee, the last farm inside the city, called the Community Food Center. Our guide is Will Allen, a Milwaukee native and once a professional basketball player, who founded and now directs Growing Power, Inc., one of the nation's most dynamic urban farming and marketing programs. Will holds forth to the rapt kids on the wonders of hydroponics, permaculture, composting, and cooperation. His recites his mantra, that Growing Power not only teaches young people to grow and eat healthy food, but it also "grows community" by connecting people to the earth and to each other (fig. 10.6).

This home base for Will's expanding network is a marvel of efficient site design and agro-technology. Its two acres (smaller than many suburban house lots) are

packed with six greenhouses, seven hoop houses, a kitchen, classrooms, administrative offices, outdoor livestock and compost areas, an apiary, a rainwater catch-
ment system, and a retail store offering produce, meat, worm castings, and compost to the community.[23] Will leads the mesmerized children, teachers, and me
through the farm in a ritual shared by some five thousand youths and adults annually. When Will invites the kids to hold worms from the vermiculture bed in their
hands, they respond with appropriate shrieks of horror. Will then casts his line
into the fish tank and hooks (in a well-rehearsed routine) a sturdy tilapia fed on
poultry wastes. A rainwater storage system irrigates long racks of vegetables and
herbs under grow lights.

As with its counterpart programs in cities across the nation, Growing Power
provides youth training and green-jobs skills through the medium of food production and marketing, while also teaching healthy diet and social skills such as
partnering, following through on commitments, and respect for others. A press
release announcing Will Allen's NEA Foundation Award for Outstanding Service
to Public Education also describes his Growing Power involvement: "Widely considered one of the leading authorities in the expanding field of urban agriculture,

Fig. 10.6. Will Allen at his Growing Power greenhouse complex in Milwaukee.

Allen teaches inner-city youth about farming, business management and marketing, by taking them through the entire process, from planting seeds to selling produce at farmers' markets. To date, he has developed partnerships with more than 10 Milwaukee Public School (MPS) schools to put into action school-based food projects that include curriculum-based programs complying with Wisconsin State Standards. Allen's organization, Growing Power, has also supplied 40,000 Milwaukee Public School children in 75 elementary schools with the food it grows."[24]

Growing Power now operates several satellite farm sites in Milwaukee and Chicago. The latter program directed by Will's daughter Erica Allen, which is based at Iron Street Urban Farm on the Near South Side, mirrors the range of agricultural facilities and training programs pioneered in Milwaukee. Other Chicago Growing Power sites include the Chicago Lights Urban Farm Program at the Cabrini-Green housing project on the North Side, and a showcase Art on the Farm urban garden in Grant Park.

Growing Power is a powerhouse within the ever-expanding urban garden and marketing movement, nationally and around the world. Will Allen has been honored many times, most notably with a MacArthur "Genius Grant," and was listed by *Time* in 2010 as one of the one hundred people "who most affect our world." Among his new projects is a proposed five-level greenhouse in Milwaukee, which he describes as the tallest "vertical farm in the world."[25]

Portland's Oaks Bottom: Rescued by a Ruse

On a bright sunny day in 2005, three environmental veterans of Portland, Oregon— Michael Houck, Gary Rideout, and Steve Johnson—show me the view from a bluff overlooking Oaks Bottom Wildlife Refuge, Portland's first officially designated urban wildlife refuge. The city skyline looms like the Emerald City beyond a panorama of open water, swamp, and forest. A great blue heron—perennial slough resident and Portland icon (honored as the name of a favorite city-brewed ale)— stands in the foreground. In the background, a seventy-foot-high mural of the same bird embellishes the side of the Portland Memorial Mausoleum, perhaps suggesting that heaven is the ultimate refuge for the true bird lover. Since then, the heron has been joined by a menagerie of birds and other Bottoms creatures in a wetland motif measuring 55,000 square feet—the largest hand-painted building mural in North America[26] (fig. 10.7).

The slough is in fact a birder's earthly paradise according to Houck, its long-time protector and champion: "If you look north across the open water [from the beginning of the loop trail] you'll see as many as 50 or sometimes 100 herons feeding, resting, or flying about, depending on the time of year. Great egrets can often be seen feeding with the herons, Common and hooded mergansers, double-

Fig. 10.7. Heron mural on crematorium overlooking Oaks Bottom Urban Wildlife Refuge, Portland, Oregon. Courtesy Michael Houck.

created cormorants, northern shovelers, mallards, green-winged teal and wood ducks are usually quite abundant. More than 100 species of birds have been seen in the bottoms."[27]

Oaks Bottom Slough is probably the first urban wildlife preserve to originate in subterfuge. By the 1980s, this wetland near the confluence of Johnson Creek and the Willamette River close to downtown Portland, was in process of obliteration. Rubble from freeway construction was deposited in the north end of the wetland. The south end was long used as a garbage dump, and fill for a railroad trestle obstructed water exchange between the slough and the Willamette River. In the early 1970s, the city proposed to drain the rest of the slough for a motocross course and other recreational purposes. According to Houck, this plan transformed Oaks Bottom into "an instant *cause celebre* and the city's first foray into debates about . . . the larger question of the city's responsibility for retaining wildlife and wildlife habitat in the urban core."[28]

In the early 1980s, after failing to persuade the city to designate the (city-owned) slough as a wildlife refuge, Houck and his co-conspirators took matters into their own hands. They created some forty brilliant yellow "wildlife refuge" notices (that were in fact modified Oregon Department of Fish and Wildlife signs that Houck requisitioned from a state wildlife biologist), and posted them around the perimeter of the slough. Finally in 1988, the city ratified this prank by adopting the Oaks Bottom Wildlife Refuge Management Plan that Houck and three colleagues drafted designating the area as a 163-acre urban wildlife refuge.[29] After serving as a catalyst for change, Oaks Bottom is "now one of Portland's urban greenspace flagships."[30]

The unorthodox rescue of Oaks Bottom helped to energize the broader movement to conserve, connect, and manage a regional system of greenspaces in metropolitan Portland—a process in which my three hosts, since that sunny day in 2005, have been key movers and shakers. The Bottom itself—now bordered by trails and connected to the city's 40-Mile Loop Trail System and its Springwater Corridor bike-pedestrian trail along the Willamette—helped to catalyze a complex planning process for the adjacent fifty-four-square-mile Johnson Creek watershed.

Johnson Creek, like its counterparts discussed in the chapter 9, was a highly developed, flood-prone, and biologically impaired urban stream by the 1990s. Flooding and fish restoration loomed large as issues that drove decades of fruitless planning for the watershed by the Army Corps of Engineers, Portland Metro, and other agencies. With the repeated failure of top-down planning to achieve results, an alternative bottom-up approach began to emerge through the efforts of local activists, particularly Steven Johnson whose family had lived and farmed along the creek for four generations. According to the Johnson Creek Watershed Council CEO Michelle Bussard: "Throughout the nineties, Steve and many others quietly amassed a coalition of neighborhood conservationists, bureaucrats, and coalition builders. From this unlikely union grew the 'marching band' that eddied into Friends of Johnson Creek, eventually plunging into the Johnson Creek Watershed Council. Johnson Creek now had the voice it had lacked."[31]

In 1996, the year that the watershed council was formed, Johnson Creek incurred its flood of record. Its newly found "voice" was joined by U.S. Representative Earl Blumenauer in procuring hazard mitigation funds from the Federal Emergency Management Agency (FEMA) to acquire selected properties in the floodplain. One of these project sites, after three years of winning over the local neighborhood, would become Brookside Wetland, a twenty-acre greenspace acquired with FEMA funds and used for recreation, wildlife habitat, and occasional storage of Johnson Creek floodwaters.

In 1998 and 1999, Representative Blumenauer and the council convened two Johnson Creek "Watershed Summits" involving public agencies, schools, non-profits, and private citizens. A "vision statement" adopted at the second summit declared: "The Johnson Creek basin will become a healthy, safe, and vibrant watershed by effectively planning for and managing growth, promoting sustainable economic development, and respecting and enhancing the natural functions and benefits of the creek."[32]

This statement reflects an emerging new perspective of ecosystem management as applied to the entire watershed, in contrast to earlier plans concerned only with mainstem flood control. It laid the philosophical foundation for the Johnson Creek Restoration Plan of 2001, which in turn led to the 2004 Johnson Creek Action Plan and the council's Five-Year Strategic Plan issued in May 2012.[33] Pursuant to this bottom up planning process, dozens of site-specific restoration projects have been accomplished in the watershed, along with the extension of the regional bikeway system, and facilities for environmental education and outdoor exercise.

Beyond The Gates: Outreach by Museums and Botanic Gardens

One of the most promising trends in humane urbanism is the increased involve-ment of "blue stocking" urban museums and botanical gardens with "blue col-lar" neighborhoods in their cities. The American Museum of Natural History in New York hosts the Center for Biodiversity and Conservation,[34] which applies the scientific expertise of the museum to conservation issues of the New York–New Jersey metropolitan region. The Field Museum in Chicago provides offices and staff in support of the Chicago Region Biodiversity Council (aka "Chicago Wilderness"),[35] as discussed later in the chapter. The Cleveland Museum of Natu-ral History now houses GreenCityBlueLake[36]—(formerly Ecocity Cleveland)—a dynamic NGO that promotes affordable housing, mass transit, energy conserva-tion, and urban farming in northeastern Ohio. Culture Connects All—a program of Partners for Livable Cities funded by MetLife and the Ford Foundation—has mobilized art museums in Chicago, New York, Tampa, Atlanta, Phoenix, and Dallas to celebrate and nurture the visual and performance arts of their ethnic and minority communities.[37]

Many urban botanical gardens now are strongly engaged with their surround-ing communities.[38] Such institutions have long served as "emerald oases" within the larger urban fabric. Their exquisite gardens and conservatories harbor micro-cosms of global ecosystems, including those native to the bioregion in which they are located. They are centers of ecological research and teaching, and they welcome large numbers of public visitors, including endless busloads of school children and teachers. Thus traditional botanical gardens are intentionally, and happily, a world apart from the turbulence of the city outside their gates.

The Brooklyn Botanic Garden (BBG) is a pioneer among its peers in respond-ing to the needs of its surrounding community. As early as 1914 when the BBG itself was founded, Ellen Eddy Shaw started its famous Children's Garden, which

Fig. 10.8. Neighbors with staff of the Brooklyn Botanic Garden GreenBridge Program. Courtesy BBG.

today draws some eight hundred children a year to tend plots and share planting know-how.

Beginning in the 1990s, BBG began to revisualize how it could better serve the vast Borough of Brooklyn (whose population of 2.5 million would make it the nation's fourth-largest city if it were still independent). In addition to welcoming the public to visit its fifty-two-acre "oasis" of gardens and conservatories, the BBG also began to "take the garden to the people." As founded in 1993 by Ellen Kirby and directed today by Robin Simmen, GreenBridge oversees a suite of outreach programs including Making Brooklyn Bloom, a free annual symposium on community horticulture; the GreenBridge Community Garden Alliance (fig. 10.8); the Greenest Block in Brooklyn Contest; the Street Tree Stewardship

Initiative (in support of the city's MillionTreeNYC planting program now in progress); the Brooklyn Urban Gardener certificate program (BUG); composting assistance and training as host of the NYC Compost Project at BBG, funded by the NYC Department of Sanitation. In addition GreenBridge contributes to BBG's partnership with the Brooklyn Academy of Science and Environment (BASE), a specialized public high school located near the garden.

Botanical institutions in Philadelphia, Chicago, Columbus, St. Louis, and elsewhere have joined BBG in creating their own outreach programs to better serve urban neighborhoods. These programs take many forms and involve complex partnerships with city and state governments, foundations, educational institutions, and local community organizations. Some involve considerable staff time and hands-on interaction with community members. Others may focus on "training the trainers" to establish a cadre of volunteers, teachers, and local citizens who in turn work with local community members on community gardens, tree planting and maintenance, invasives removal, and other urban botanical activities. Such efforts resonate with broader urban goals, including:

- Food security—ensuring availability of affordable, nutritious food.
- Healthy diet, physical fitness, and combating obesity.
- Youth training in growing and marketing food ("green jobs").
- Urban beautification such as street tree planting and neighborhood floral displays.
- Reclamation of vacant lots for gardens and recreation.
- Environmental justice.
- Environmental education.
- Community pride and neighborliness.

Community-based garden programs in particular connect botanical gardens with the urban agriculture movement stimulated by such catalysts as Growing Power (Milwaukee and Chicago), Edible Schoolyard (Berkeley), Green Thumb and Added Value (New York City), Nuestras Raices (Holyoke, Massachusetts), and their counterparts elsewhere. Botanical gardens may reinforce and enrich the broader urban gardening movement by contributing their research and teaching expertise to the mix of urban farming programs. They are particularly suited to providing technical training in such areas as composting, plant selection, organic practices, seed banks, rain gardens, horticultural therapy and so forth. They also may serve as "honest brokers" to help allocate governmental or foundation funds to deserving neighborhood programs and monitor the success of local planting initiatives over time. Several flagship programs are summarized in BBG's recent publication, *Community Gardening* (Ellen Kirby and Elizabeth Peters, editors, 2008).

Tying It All Together: Roles of Regional Networks

Grassroots initiatives like those discussed in this chapter usually operate on shoe-string budgets and in relative isolation from others that perform similar work. An important vehicle for connecting programs and people engaged in kindred issues and activism since the 1990s has been the formation of regional alliances of government agencies, corporations, environmental organizations, and individuals to pursue shared objectives. Five exemplars around the country worthy of mention here include the New York Metropolitan Waterfront Alliance (MWA), the Lake Erie Partnership for Biodiversity (LEAP), Chicago Wilderness, Houston Wilderness, and the Intertwine Alliance in metropolitan Portland, Oregon.

Alliances take various forms, legally and organizationally. Some are incorporated like New York's Metropolitan Waterfront Alliance. Others are unincorporated networks like Chicago Wilderness, LEAP, and the Intertwine. Some charge annual fees and solicit contributions from members; others seek to optimize their membership by charging nominal or no fees. Most have by-laws, staff, working groups, and periodic membership meetings or conferences. Most depend on government and foundation financial support.

Metropolitan Waterfront Alliance

With over 620 member organizations, the Metropolitan Waterfront Alliance (MWA) based in New York City is probably the nation's largest environmental network within a single urban region. MWA's mission is "to transform the waters of New York and New Jersey Harbor into clean and accessible places to learn, work and play, with inviting parks, dependable jobs and reliable, eco-friendly transportation for all."[39] It was founded in the early 2000s as a program of the New York Municipal Art Society, a leading civic institution. Under the leadership of Roland Lewis (formerly the head of New York Habitat for Humanity), and education director Cortney Worrall, MWA expanded rapidly in membership and areas of concern. In 2007 it incorporated as a nonprofit and moved to a new and suitably maritime office location at the South Street Seaport in the shadow of the Brooklyn Bridge. From there it exerts considerable influence in the gradual transformation of New York's long-degraded waterfronts for new uses and users—a vast exercise in humane urbanism.[40]

The program of MWA, as organized in its 2008 "Waterfront Agenda,"[41] operates through six task forces drawn from the ranks of its member organizations:

1. Working Waterfront—reviving harbor-related commercial activity.
2. Mass Water Transit—promoting wider use of ferries and water taxis.
3. Green Harbor—supporting research and restoration of harbor ecosystems.

4. Harbor Education—strengthening harbor-based training and education programs.
5. Harbor Recreation—revitalizing waterfront access and water-related activities.
6. Aquatecture—appropriate design of physical waterfront facilities.

Serving one of the world's largest, most complex urban regions, MWA has already established itself as a major waterfront voice and catalyst. The city's 2011 Comprehensive Waterfront Plan, an element of Mayor Michael Bloomberg's *PlaNYC,* was strongly influenced by the alliance and its task forces. Its 2011 conference (modestly priced) attracted hundreds of waterfront citizens and professionals for paper sessions, meetings, and a harbor cruise at sunset. With MWA's help, New York has much to teach twenty-first-century megacities worldwide about how to sustain nearly twenty million contentious people and their degraded but newly appreciated natural environment.[42]

Chicago Wilderness

The Chicago Biodiversity Council, aka Chicago Wilderness (CW), was established in the mid-1990s to promote biodiversity and environmental education in greater Chicago, including southeastern Wisconsin and northern Indiana.[43] CW is an open-access, public-private consortium of regional stakeholders concerned with the protection, restoration, and management of habitat sites, as well as research and education on biodiversity. While CW does not per se take positions on biodiversity issues, its value lies in facilitating collaborative efforts to analyze issues and formulate recommendations for public policy by subgroups of member organizations organized as task forces.

CW currently includes about three hundred member governmental agencies, NGOs, educational institutions, and business corporations. Its geographic reach loosely includes the six Illinois counties of the Chicago Metropolitan Statistical Area, Kenosha County in Wisconsin, and Lake and Porter counties in Indiana. Office space and staff resources are provided by three Chicago-area organizations: the Field Museum of Natural History, the Brookfield Zoo, and The Nature Conservancy Chicago Chapter. Startup funding was provided by grants from the U.S. Environmental Protection Agency, the U.S. Forest Service, and the U.S. Fish and Wildlife Service for research on biodiversity.

Chicago Wilderness—a national model for urban biodiversity advocacy—is "governed" by three leadership entities established under its Policies and Procedures: (1) an Executive Council comprising the above three organizations plus additional members that provide resources to CW; (2) a Steering Committee that includes representation of specified sectors and classes of governments and private interest groups; and (3) a Coordinating Group established by the Steering

Committee, which holds monthly meetings open to all CW members. The coordinating group implements steering committee decisions, oversees the CW work plan, sets agendas for meetings, and represents CW at professional meetings.

Unlike New York's MWA, Chicago Wilderness is not incorporated and does not have tax-exempt status so that it does not compete with its member organizations for funding. The work of CW is carried out through meetings of members, mission-specific task forces, a proposals committee, and a nominating committee. The CW Corporate Council includes participating business firms. CW supports ecological restoration activities through a network of citizen volunteers.

In a decade and a half, Chicago Wilderness has become a respected voice for "ecological citizenship" in the Chicago region. In 1999–2000, it published its profusely illustrated *Atlas of Biodiversity* and *Biodiversity Recovery Plan* which document respectively the "what" and "how" of ecological stewardship in the region. CW's "No Child Left Inside" program coordinates environmental education programs for inner city and suburban school systems. Member institutions like the Field Museum of Natural History, the Morton Arboretum, and the U.S. Fish and Wildlife Service conduct research on biodiversity through CW task forces and subgroups. CW also participates in regional initiatives such as the Chicago Regional Transportation Plan and the Green Infrastructure Regional Mapping Project.

Lake Erie Allegheny Partnership for Biodiversity

Among a handful of urban biodiversity alliances inspired by Chicago Wilderness elsewhere, the Lake Erie Allegheny Partnership for Biodiversity (LEAP), was formed in 2004 as a public-private partnership in the Greater Cleveland region. Geographically, it embraces "the glaciated lands and waters south of Canada from Sandusky Bay to the Allegheny Mountains"—an area of some seventeen thousand square miles extending across northeastern Ohio, northwestern Pennsylvania, and western New York. As of early 2012, the LEAP alliance totaled about fifty member organizations including government agencies, universities, and conservation organizations engaged in promoting biodiversity, environmental education, and public outreach. The partnership strives to initiate research and share technical information, conduct public education and outreach efforts, and establish core volunteer groups.[44]

Like Chicago Wilderness (and unlike New York's MWA), LEAP focuses on the single issue of protecting and interpreting regional biodiversity. Its membership includes major state and regional parks and conservation agencies, land trusts, and other land-holding organizations. Collectively, LEAP's members own and manage some 204,237 acres with another 37,591 acres under protective easements. The LEAP Regional Biodiversity Plan Committee is developing a BioMap model developed by Massachusetts Department of Fish and Game and The Nature Conservancy.[45]

Unlike the other alliances discussed here, LEAP does not have a corporate council or major public support. It is a spin-off of Cleveland's respected environmental advocacy group GreenCityBlueLake (formerly EcoCity Cleveland), which is a subsidiary of the Cleveland Museum of Natural History.

Houston Wilderness

Another Chicago Wilderness spin-off is Houston Wilderness, which defines itself in its mission statement as "a broad-based alliance of business, environmental and government interests that act in concert to protect, preserve and promote the unique biodiversity of the region's precious remaining ecological capital—from bottomland hardwoods and prairie grasslands to pine forests and wetlands—while recognizing the importance of the region's natural assets to its cultural history, economic vitality and future well-being.[46]

Houston Wilderness (HW) serves a huge geographical region, including the ten-county Greater Houston Metropolitan Region—the nation's sixth-largest metro area with 5.6 million inhabitants—plus fourteen surrounding counties. Its list of seventy-plus member organizations includes many area governments, business corporations, and educational organizations, as well as the Buffalo Bayou Partnership and the Bayou Preservation Association discussed in chapter 9. HW provides various educational and recreational activities to connect Houstonians with nature in their communities and region. Research and restoration are currently less a priority than with its Chicago counterpart.

Greater Houston, surprisingly to non-Texans perhaps, is a very biodiverse region with ten distinct eco-regions including pine forests, savanna, prairie, river bottomlands, coastal salt marshes. All ten of its eco-habitats will be linked by the proposed "Sam Houston Greenbelt," a vast chain of greenspaces and trails encircling Houston and connecting existing parks, refuges, beaches, and waterways—the signature undertaking of Houston Wilderness and its partners.

The Portland Intertwine Alliance

In metropolitan Portland, Oregon, the creation of the Oaks Bottom Urban Wildlife Refuge discussed earlier in the chapter energized efforts in the 1980s to protect and connect natural areas throughout the Portland–Vancouver, Washington region. A major outgrowth of Portland's strong sense of regional identity and stewardship was the passage of a $227.4 million open space acquisition bond issue in 2006. Helping to get that adopted was a new ad hoc partnership, the nucleus of what would become The Intertwine Alliance (TIA). TIA coalesced under the leadership of Portland Metro, the only elected regional government in the nation, in collaboration with Trust for Public Land, Portland Audubon, and the Urban Greenspaces Institute. A regional summit conference in 2007 (at which I was a speaker) helped to launch the alliance as a partnership of area governments, park and natural resource agencies, colleges, and environmental organizations.

In 2011, TIA became a nonprofit corporation called The Intertwine Alliance Foundation. Its list of contributing partner organizations has grown rapidly, from twelve in 2009–10, to fifty in 2011–12. Another twelve partners have been welcomed without providing a financial contribution. A core group among its directors oversees the administration of the alliance. An executive director is now in place and additional staff will be added as funding allows.

The mission of TIA broadly addresses five priorities: (1) conservation; (2) environmental education; (3) open space acquisition; (4) active transportation (cycling jogging, walking, hiking); and (5) linking the regional open space system. Its early activities have included developing a uniform signage for use along the region's trails and greenways, promotion of conservation, education, and recreation through the activities of its partners, and developing public awareness of the "Intertwine" as the nexus of Portland's sense of place, natural resources, and lifestyle choices.[47]

Epilogue

Democratized by necessity, the [Fort Collins long-range planning] process led to goals that went beyond the predictable safe streets and commerce that planners might have otherwise emerged. In a departure from the old command-down process—planners proposing, residents disposing in public planning meetings—ideas bubbled up in new ferment.

—KIRK JOHNSON, "A Town Envisions the Future on Its Own Terms," *New York Times,* November 17, 2011

The spectrum of humane urbanism across the country is broad and open-ended, defined as it is by local ingenuity—"ideas bubbling up in new ferment"—instead of top-down fiat. Humane urbanism eschews grand plans, textbook designs, and mega-development that breeds gentrification. Its aesthetics evolve not from established standards of architectural and planning design, but from the spontaneous palettes of mural artists, urban gardeners, building renovators, the melee of street fairs and ethnic festivals, and the rainbow of people—diverse in age, race, life style, wealth, and apparel—who share urban spaces and experience.

Grassroots efforts to make communities more "humane" are often scattered, uneven, and underfunded. They are also hard to assign to neat classifications. In the introduction, I likened humane urbanism to *topsoil*—a nurturing medium that promotes creative symbiosis among people, places, and plants. And in the preface I suggested that the collective efforts of local movers and shakers comprise the *compost* of humane urbanism, enriching the fertility of that topsoil.

Synergy among diverse goals is a critical attribute of humane urbanism. Society tends to divide its attention and resources among competing needs—jobs, housing, education, health, and environment—and then address each one separately, if at all. But at the community scale (however defined), these needs must be confronted simultaneously or progress in one area will be undermined by failure in others. That of course is the modus operandi of community development corporations and their national umbrellas like the Local Initiative Support Corporation (LISC), as discussed in chapter 7.

Like ecology, humane urbanism thrives on diversity—of participants, of goals, of means, of disciplines, and (one hopes) of viewpoints. Some initiatives are closely related to larger national movements—social and environmental justice, afford-able housing, school choice, health and fitness, natural disaster mitigation, animal rights, and climate change mitigation. Others are truly homegrown, as in saving the Forbus Butternut in Poughkeepsie or protecting the Oaks Bottom Slough in Portland, Oregon. They depend on spontaneous and often voluntary local leader-ship. They are pragmatic and creative in stitching together existing program re-sources, available funding, and donations of money, time, and office space. Most involve public-private partnerships, some of which are local alliances to save a par-ticular site, to restore a stream, wetland, or watershed, or pursue a particular mis-sion such as environmental education or urban gardening. Others have evolved into influential regional networks such as Chicago Wilderness. Many also foster social interaction among diverse populations sharing a common resource like a watershed, thus promoting what the ethicist Andrew Light refers to as "ecological citizenship."[1]

Humane urbanism is not a panacea or "silver bullet" solution to the metropoli-tan muddle created over the past century. Most freeways will not be replaced by light rail and bikeways, and will certainly become more crowded and deteriorated in coming years. Urban greenhouse gases will continue to warm the earth's atmo-sphere despite the plethora of climate change plans adopted by cities associated with the International Council for Local Government Initiatives (ICLEI). Urban farming programs such as Growing Power will provide increasing supplies of fresh, healthy, locally grown food, but junk food will not disappear anytime soon. Affordable housing will continue to be scarce and remote from employment op-portunities despite the dedicated work of such housing providers as LISC, Habitat for Humanity, and Enterprise Community Partners.

This book does not attempt to predict the future path of American cities; that it leaves to others whose crystal balls are clearer than mine. Nor does it seek to prescribe any universal elements of humane urbanism. That task is left to the good sense of those engaged in the process who best know what they need and what resources may be available, or not. The purpose of this book, as I stated in the preface, has been to explore "the terrain of humane urbanism, both in its present contours as far as they can be discerned and historically in its relationship to what has gone before." We have crossed a major hurdle in overturning the hegemony of patricians and technocrats—their help is of course still welcome, but the goals and means from now on must be increasingly determined by those who will live with the results.

Notes

Preface

1. White was awarded the National Medal of Science by President Clinton and the National Geographic Society's highest award, the Hubbard Medal, among other tributes to his work on water resources, natural hazards, and the environment.

2. William H. Whyte, *The Last Landscape* (Garden City, NY: Doubleday, 1968); the book was republished by the University of Pennsylvania Press in 2002.

Introduction

1. Boston & Maine Railroad, "Valley of the Connecticut and Northern Vermont" (1901), reprinted in W. D. Wetherell, ed., *This American River: Five Centuries of Writing about the Connecticut* (Hanover, NH: University Press of New England, 2002).

2. Bill McKibben, "The Conte Refuge" (1995), reprinted ibid., 287.

3. William Cronon, *Changes in the Land: Indians, Colonists, and the Ecology of New England* (New York: Hill & Wang, 1983). Old riverside town names today recall the importance of the natural meadows of the Connecticut River lowlands to early English settlers: Springfield, Longmeadow, Enfield, Suffield, Westfield, and Northfield. (The "field" place-name tradition was continued in the 1960s with construction of "Eastfield Mall.")

4. The gradual displacement of Native populations by the colonists generated much suffering on both sides and occasional outright conflict, most notably the 1675 Native rebellion known as King Philip's War.

5. Cronon, *Changes in the Land,* 33.

6. Paul Zielbauer, "Poverty in a Land of Plenty: Can Hartford Ever Recover?" *New York Times,* August 26, 2002.

7. Catherine Tumber, *Small, Gritty, and Green: The Promise of America's Smaller Industrial Cities in a Low-Carbon World* (Cambridge: MIT Press, 2012), 28.

8. Alan Mallach, ed., *Rebuilding America's Legacy Cities: New Directions for the Industrial Heartland* (Washington, DC: Brookings Institution, 2012); for information on Park River Initiative, see www.parkwatershed.org/about-us/.

9. Information available at www.nuestras-raices.org/; Tumber, *Small, Gritty, and Green,* 85–87.

10. Joel Garreau, *Edge City: Life on the New Frontier* (New York: Doubleday Anchor, 1991).

11. According to Wikipedia, Co-op City today is 55% African American, 25% Hispanic and 20% non-Hispanic white.

12. Paul S. Grogan and Tony Proscio, *Comeback Cities A Blueprint for Urban Neighborhood Revival* (Boulder, CO: Westview Press, 2000).

13. New York City Department of Planning, *PlaNYC: A Greener, Greater New York,* 2007, www.nyc.gov/html/planyc2030/html/theplan/the-plan.shtml.

14. Michael Kimmelman, "River of Hope in the Bronx," *New York Times,* July 12, 2012.

15. The editors of the *New York Times* wryly observed: "Any city gets what it admires, will pay for, and, ultimately, deserves. Even when we had Penn Station, we couldn't afford to keep it clean. We want and deserve tin-can architecture in a tinhorn culture. And we will probably be judged not by the monuments we build but by those we have destroyed." Editorial, "Farewell to Penn Station," *New York Times,* October 30, 1963.

16. As I recount in the preface, the concept of humane urbanism was loosely derived from the work of William H. Whyte, memorialized in the 2002 Humane Metropolis Conference in New York City, which was supported by the Lincoln Institute of Land Policy, Laurence S. Rockefeller, and other sources. That conference led to the volume *The Humane Metropolis: People and Nature in the 21st-Century City* (Amherst: University of Massachusetts Press, 2006), which I edited and which in turn was the template for three additional Humane Metropolis symposia, in Pittsburgh, Riverside (California), and Baltimore, all sponsored by the Lincoln Institute.

17. Tumber, *Small, Gritty, and Green.*

18. Joel Kotkin, *The New Geography: How the Digital Revolution is Reshaping the American Landscape* (New York: Random House, 2000).

19. Don Peck, "Can the Middle Class be Saved?," *Atlantic,* September 2011, 60–78.

20. Nicholas Lemann, "Get Out of Town," *New Yorker,* June 27, 2011, 79 (emphasis added).

21. Paul S. Grogan and Tony Proscio, *Comeback Cities: A Blueprint for Urban Neighborhood Revival* (Boulder, CO: Westview Press, 2000), 101 (emphasis added).

22. Kotkin, *New Geography,* 59.

23. Jennifer Wolch, "Green Urban Worlds," *Annals of the Association of American Geographers* 97.2 (2007): 373–84, quot. 374.

1. American Cities in 1900

1. In 1900, approximately 60% of the U.S. population (45.4 million) lived on farms or in villages, while the other 40% (30.4 million) was classified as "urban."

2. Erik Larson, *The Devil in the White City: Murder, Magic, and Madness at the Fair That Changed America* (New York: Random House, 2003).

3. Jon C. Teaford, *The Twentieth-Century American City,* 2nd ed. (Baltimore: Johns Hopkins University Press, 1993), 7.

4. Frederick Jackson Turner, "The Significance of the Frontier in American History" (1893), reprinted in *City and Country in America,* ed. David R. Weimer, 62–76 (New York: Appleton-Century-Crofts, 1962), 74–75.

5. Edmund Morris, *Theodore Rex* (New York, Random House, 2001), 20.

6. After the loss of the World Trade Center in 2001, Chicago's Sears Tower (now Willis Tower) became the tallest skyscraper in the United States at 1,450 feet. At this writing, eight even taller buildings now stand in Dubai, Mecca, Taipei, Shanghai, Hong Kong, and Kuala Lumpur (Petronas Towers 1 and 2).

7. Libby Hill, *The Chicago River: A Natural and Unnatural History* (Chicago: Lake Claremont Press, 2000).

8. The original Croton system, which opened in 1842 and expanded in the 1890s, inspired Boston and many other cities to follow New York's strategy of diverting fresh water from upland sources and delivering it by gravity flow to the user region with minimal treatment. The New York City water system today serves ten million people, with total demand averaging one billion gallons per day, mostly from its newer reservoirs west of the Hudson

River in the Catskill Mountains and upper Delaware River basin. See National Research Council, *Watershed Management for Potable Water Supply: Assessing the New York Strategy* (Washington, DC: National Academy Press, 2000); Rutherford H. Platt, Paul K. Barten, and Max J. Pfeffer, "A Full, Clean Glass: Managing New York City's Watersheds," *Environment* 42.5 (2000): 8–19.

9. Many of those who entered the caissons, including Washington Roebling, were stricken with life-threatening onset of "the bends" owing to the poorly understood effects of depressurization on the human body.

10. David McCullough, *The Great Bridge: The Epic Story of the Building of the Brooklyn Bridge* (New York: Simon & Schuster, 1972).

11. Edwin G. Burrows and Mike Wallace, *Gotham: A History of New York City to 1898* (New York: Oxford University Press, 1999), 1235.

12. Neal Peirce and Curtis W. Johnson, *Century of the City: No Time to Lose* (New York: The Rockefeller Foundation, 2008), 26.

13. Eric Homberger, *The Historical Atlas of New York City* (New York: Henry Holt, 2005), 105.

14. The Twentieth Century Limited continued in service until 1967. The train was commemorated in many movies, most notably as the setting of the rolling crap game in *The Sting*.

15. Sue Kohler and Pamela Scott, eds., *Designing the Nation's Capital: The 1901 Plan for Washington, D.C.* (Washington, DC: U.S. Commission of Fine Arts, 2006).

16. Charles A. Keeler, *San Francisco and Thereabout* (San Francisco: California Promotion Committee, 1902), 94.

17. For further discussion see Rutherford H. Platt, *Disasters and Democracy: The Politics of Extreme Natural Events* (Washington DC: Island Press, 1999), 246. The recovery of San Francisco was a mixed legacy. It catalyzed the development of a new source of water from the Sierra Nevada 150 miles east of the city to control future fires. However, like London after its Great Fire of 1666, San Francisco did not alter its basic layout of streets and land use patterns as it rebuilt, even ignoring Daniel H. Burnham's City Beautiful plan for its redesign, which had been presented to civic leaders just before the fire.

18. Catherine Tumber, *Small, Gritty, and Green: The Promise of America's Smaller Industrial Cities in a Low-Carbon World* (Cambridge: MIT Press, 2012), xxiii.

19. Harvey K. Flad and Clyde Griffen, *Main Street to Main Frames: Landscape and Social Change in Poughkeepsie* (Albany: SUNY Press of Albany, 2009), 81.

20. Ibid., 63, 71.

21. Sam Bass Warner Jr., *Streetcar Suburbs: The Process of Growth in Boston (1870–1900)* (Cambridge: Harvard University Press, 1962), 157.

22. Richard Hofstadter, *The Age of Reform* (New York: Knopf, 1955), 5.

23. Warner, *Streetcar Suburbs*, 112–13.

24. Kenneth T. Jackson, *Crabgrass Frontier: The Suburbanization of the United States* (New York: Oxford University Press, 1985), 146–48.

25. Ibid. As discussed later, the urban sociologist William H. Whyte in *The Organization Man* (Garden City, NY: Doubleday, 1956) documented much the same homogeneity among residents of the new postwar planned suburb of Park Forest, Illinois.

26. Homberger, *The Historical Atlas of New York City,* 110.

27. Jacob A. Riis. *How the Other Half Lives: Studies among the Tenements of New York, 1890,* ed. Sam Bass Warner (Cambridge: Belknap Press of Harvard University Press, 2010; originally published 1890).

28. Quoted in Tyler Anbinder, *Five Points: The 19th-Century New York City Neighborhood That Invented Tap Dance, Stole Elections, and Became the World's Most Notorious Slum* (New York: Free Press, 2001), 358.

29. Lawrence Veiller, "The Tenement-House Exhibition of 1899," *Charities Review* 10: 19–25, quotation on 19. Available at http://tenant.net/community/Les/viller1.html.

30. That law and its 1866 predecessor, the New York Metropolitan Health Act (a state law) were motivated by a prevailing belief in the "miasma theory" of disease, namely that cholera, tuberculosis, and other scourges of slum districts were caused by the presence of filth and bad air. The theory was most famously propounded by the English sanitary reformer Edwin Chadwick, whose work influenced the first sanitary survey in New York City by Dr. John Griscom, author of *The Sanitary Condition of the Laboring Population of New York* (1845). The miasma theory was eventually replaced with a more scientific understanding of microbiological factors in infectious diseases.

31. Burrows and Wallace, *Gotham*, 1173. Shortly after the Veiller Exhibition, the Tenement Act was amended to outlaw airshafts as the sole source of daylight and ventilation; thenceforth New York required that rooms in new buildings have windows opening to a street or rear courtyard, as well as a toilet for each household unit, fire escapes, and hallway lighting. Many buildings of that era still stand today in various stages of decay and renovation.

32. Upton Sinclair, *The Jungle*, (1906; New York: MUF Books 2003), 20. According to Sinclair, it was not uncommon for workers in meatpacking plants to fall into the cauldrons of boiling fat and thus be reduced to lard.

33. Ibid., 7.

34. The 1889 Johnstown flood described later in this chapter literally reflected this geographic relationship: over 2,200 steel workers and their families died when an upstream dam owned by a club of steel executives collapsed.

35. Morris, *Theodore Rex*, 47.

36. Ibid.

37. Raymond Arsenault, *The Sound of Freedom: Marian Anderson, the Lincoln Memorial, and the Concert That Awakened America* (New York, Bloomsbury Press. 2009), 17, 104.

38. Charles R. Morris. *The Tycoons* (New York: Henry Holt, 2005).

39. David G. McCullough, *The Johnstown Flood* (New York: Simon and Schuster, 1968).

40. Warner, *Streetcar Suburbs*, 155.

41. Roger Biles, *The Fate of the Cities: Urban America and the Federal Government, 1945–2000* (Lawrence: University Press of Kansas, 2011), 1.

42. For further discussion see Rutherford H. Platt, *Land Use and Society: Geography, Law, and Public Policy,* rev. ed. (Washington, DC: Island Press, 2004), chap. 4.

43. Ibid., 82–84.

44. Burrows and Wallace, *Gotham*, 784–86.

2. Competing Visions in the Progressive Era

1. Richard Hofstadter, *The Age of Reform* (New York: Knopf, 1955), 5.

2. Erik Larson, *The Devil in the White City: Murder, Magic, and Madness at the Fair that Changed America* (New York: Random House, 2004).

3. Charles Mulford Robinson, *Modern Civic Art: or The City Made Beautiful* (New York: Putnam, 1903), 1.

4. William H. Wilson, *The City Beautiful Movement* (Baltimore: Johns Hopkins University Press, 1989), 46–47.

5. *The Improvement of the Park System of the District of Columbia.* S. Report No. 166 (1902).

6. Sue Kohler and Pamela Scott, *Designing the Nation's Capital: The 1901 Plan for Washington, D.C.* (Washington: U.S. Commission of Fine Arts, 2006), 11–13.

7. Carl Smith, *The Plan of Chicago and the Remaking of the American City* (Chicago: University of Chicago Press, 2006), 66–67.

8. Daniel H. Burnham and Edward H. Bennett, *Plan of Chicago* (Chicago: Commercial Club of Chicago, 1909), excerpt reprinted in David R. Weimer, ed., *City and Country in America* (New York: Appleton-Century-Crofts, 1962), 89.

9. Weimer, ibid., 95 and 97.

10. Janice Metzger, *What Would Jane Say? City-Building Women and a Tale of Two Chicagos* (Chicago: Lake Claremont Press, 2009), 148–49.

11. Alexander Garvin, *The American City: What Works and What Doesn't* (New York: McGraw-Hill, 2002), 512.

12. Today the District owns about 67,800 acres or about 11% of Cook County, Illinois.

13. Catherine Tumber, *Small, Gritty, and Green: The Promise of America's Smaller Industrial Cities in a Low-Carbon World* (Cambridge: MIT Press, 2012), xxvi.

14. Robert W. Rydell and Laura Bird Schiavo, eds., *Designing Tomorrow: America's World's Fairs of the 1930s* (New Haven: Yale University Press, 2010).

15. Louise W. Knight, *Jane Addams: Spirit in Action* (New York: Norton, 2010).

16. Jane Addams, *Twenty Years at Hull-House* (New York: Macmillan, 1910), 98, 99–100.

17. Edmund Morris, *Theodore Rex* (New York: Random House, 2001).

18. Addams, *Twenty Years at Hull-House*, 112.

19. Knight, *Jane Addams*, 122–23.

20. Walter Lippmann, *A Preface to Politics* (New York: Mitchell Kennerley, 1914), 152.

21. Barbara Garland Polikoff, *With One Bold Act: The Story of Jane Addams* (Chicago: Boswell Books, 1999), 124–25.

22. Quoted in Knight, *Jane Addams*, 106.

23. Addams, *Twenty Years at Hull-House*, 260.

24. Ibid., 285–86.

25. Smith, *The Plan of Chicago*, 29.

26. Ibid.

27. As I discuss in chapter 3, the federal urban renewal program was upheld by the U.S. Supreme Court in a 1954 majority opinion written by Justice William O. Douglas as a valid exercise of authority to make cities "beautiful as well as sanitary." *Berman v. Parker* 75 S.Ct. 98 (1954), at 102–3.

28. Smith, *The Plan of Chicago*, 78.

29. Ibid. (Smith notes however that, "small as her role was . . . Addams was among the very few women who had any part at all in the preparation of the *Plan*.")

30. Metzger, *What Would Jane Say?*, 107. It has come to light that draft sections on these issues were deleted from the final *Plan*, apparently to placate objections from Commercial Club members. According to Metzger (107), Burnham "may have inhabited a position in the intellectual middle between the business ethic of the time and the settlement house ethic, at least in broaching questions of education, public health, and safety."

31. Metzger, *What Would Jane Say?*, 29–30.

32. Ibid., 126. See chapter 8 for my comparison of Grant and Millennium parks. It should be noted, however, that the 1909 *Plan* envisioned expansion of the city's broader park system including additions to the lakefront parks and improvement of several existing large community parks including Garfield, Humboldt, and Douglas parks on the city's West Side.

33. Knight, *Jane Addams: Spirit in Action,* 267.

34. Jon C. Teaford, *The Twentieth-Century American City: Problem, Promise, and Reality* (Baltimore: Johns Hopkins, 1993), 31.

35. Eugenie L. Birch, "Woman-Made America: The Case of Early Public Housing Policy," in *The American Planner: Biographies and Recollections,* ed. Donald A. Krueckeberg (New York: Methuen, 1983), 149–75.

36. Benjamin Clarke Marsh, *An Introduction to City Planning* (New York: The Author, 1909). The book opens with the adage: "A City without a Plan Is Like a Ship without a Rudder."

37. Harvey A. Kantor, "Benjamin C. Marsh and the Fight over Population Congestion," in Krueckeberg, *American Planner,* 69–70, n. 39.

38. *City Planning.* Sen. Doc. No. 422 at 75 (1910).

39. Ibid.

40. Kantor, "Benjamin C. Marsh," 67. Henry George and his disciples supported a tax on increases in land value, separate from buildings. The idea was to encourage landowners to put land to productive use and to recoup for the public the "unearned increment" in land value attributable to community improvements.

41. Ibid, 71.

42. City Beautiful adherents looked to France and Italy for their inspiration, whereas planning and zoning enthusiasts most admired Germany, before 1914 at least.

43. Sir William Blackstone, *Commentaries on the Laws of England,* vol., 2, bk. 2, chap. 1.

44. 260 U.S. 393 (1922).

45. Ibid. at 415; my emphasis. The lone dissent by the progressive justice Louis Brandeis argued that the law was constitutional as necessary to protect the "public health, safety, and welfare" from the effects of surface collapse.

46. Seymour Toll, *Zoned American* (New York: Grossman, 1969).

47. *Village of Euclid, Ohio, v. Ambler Realty Co.,* 297 F. 307 (N.D. Ohio, 1924) at 312.

48. *Village of Euclid, Ohio, v. Ambler Realty Co.,* 272 U.S. 365 (1926) at 370.

49. Ibid. at 395.

50. Ibid.

51. Both acts were developed at the behest of then secretary of commerce Herbert Hoover as models for state adoption, with the result that "Euclidean Zoning" evolved along very similar lines from one state to another.

52. 272 U.S., 365 at 390.

53. Thomas Adams, *Planning the New York Region* (New York: Regional Plan of New York and Environs, 1927), 35.

54. Harvey A. Kantor, "Charles Dyer Norton and the Origins of the Regional Plan of New York," in Krueckeberg, *American Planner,* 184.

55. William H. Wilson, "Moles and Skylarks" in *Introduction to Planning History in the United States,* ed. Donald A. Krueckeberg (New Brunswick, NJ: Rutgers Center for Urban Policy Research, 1983), 99.

56. The 1929 RPA Regional Plan was followed by a Second Regional Plan in 1968, and a Third Regional Plan in 1996.

57. Marion Clawson, *New Deal Planning: The NRPB* (Baltimore: Johns Hopkins University Press for Resources for the Future, 1981).

58. Birch, "Woman-Made America," 149–57.

59. Lewis Mumford, "The Plan of New York" and "The Plan of New York II," *New Republic* 71 (1932), 121–26 and 146–53, quotation on 124–25.

60. Ibid., 124.

61. Ibid., 123. While dismissing the entire enterprise as misguided, Mumford gave only cursory attention to the Plan's actual details. Amazingly from today's perspective, he endorsed the proposed construction of an elevated East River Drive to "help open up that dead area at the heel of Manhattan." He also supported a proposal to build a "mole and breakwater" between Rockaway Beach in Queens and Sandy Hook in New Jersey, which effectively would have walled off the Hudson River estuary from the Atlantic Ocean, vastly changing the ecology of the harbor. (This notion has been revived in the wake of Hurricane Sandy in 2012 by Robert Yaro, president of the Regional Plan Association, and others.) Mumford had no problem with filling in Jamaica Bay for new housing convenient to Manhattan (so much for its future National Wildlife Refuge designation). A comparable proposal to fill the Hackensack Meadows in New Jersey for industry, housing, and a regional park would remove "a natural obstacle to transportation and would possibly diminish the number of mosquitos." However, he vehemently rejected "the colossal highway and rapid transit schemes outlined by the Regional Plan [which are] an alternative to a comprehensive building program, certainly not a means to it" (ibid., 149).

62. Robert Fishman, *Urban Utopias in the Twentieth Century: Ebenezer Howard, Frank Lloyd Wright, Le Corbusier* (New York: Basic Books, 1977).

63. This concept was proposed in Howard's 1898 tract, *To-morrow: A Peaceful Path to Real Reform,* which was reissued in 1902 as *Garden Cities of To-morrow.* In his introduction to the 1965 MIT Press edition of the book, Mumford wrote: "*Garden Cities* . . . has done more than any other single book to guide the modern town-planning movement and to alter its objectives." See Ebenezer Howard, *Garden Cities of To-Morrow,* ed. F. J. Osborn (Cambridge: MIT Press, 1965), 29.

64. The Town and Country Planning Association, which evolved from Howard's Garden City Association became a leading advocate of the British new town program after World War II and continues today to advocate for decentralization.

65. Peter Hall, *Cities of Tomorrow: An Intellectual History of Urban Planning and Design in the Twentieth Century,* 3rd ed. (Oxford: Blackwell, 2002).

66. Patrick Geddes, *Cities in Evolution* (1915; New York: Oxford University Press, 1950), 38. In the habit of urbanists to invent new words to sound more profound, Geddes distinguished between what he termed the *Paleotechnic* and *Neotechnic* eras of civilization and cities. The former referred to the industrial age then still in progress with its emphasis on production and commodification of labor. The latter term described his postulated ideal future social order.

67. Ibid.

68. RPAA members included architects Clarence Stein and Henry Wright, housing advocate Catherine Bauer, and landscape architect Benton MacKaye ("the Father of the Appalachian Trail").

69. Hall, *Cities of Tomorrow,* 137.

70. Sunnyside inaugurated the "superblock"—a design concept wherein each unit opens onto a small garden or common greenspaces rather than city streets—described by Mumford as "an admirable device for lowering road costs, increasing the amount green space, and creating tranquil domestic quarters free from through traffic." Lewis Mumford, introduction to Clarence S. Stein, *Toward New Towns for America,* 3rd ed. (1957, Cambridge: MIT Press, 1966). Mumford and his wife lived at Sunnyside on Long Island from 1925 until 1936, when they moved to the patrician countryside of Duchess County, New York.

71. Eugenie L. Birch, "Radburn and the American Planning Movement," in Krueckeberg, *Introduction to Planning History,* 126.

72. Ibid.

73. Ibid., 128.

74. David Myhra, "Rexford Guy Tugwell: Initiator of America's Greenbelt New Towns, 1935–36" in Krueckeberg, *American Planner,* 225–49. President Franklin D. Roosevelt established the Resettlement Administration by executive order in 1935.

75. Hall, *Cities of Tomorrow,* 132.

76. Ibid., 103.

77. Ibid., 126.

78. Quoted ibid., 146.

79. Before the depression, automobile ownership grew by leaps and bounds in the United States with 2.3 million cars sold in 1922, more than 3 million annually from 1923 to 1926, over 3.8 million in 1927 and 1928, and 4.5 million in the fateful year of 1929. Mel Scott, *American City Planning: Since 1890* (Berkeley: University of California Press, 1971), 183.

80. Frank Lloyd Wright, "Broadacre City: A New Community Plan" (1935), reprinted in *City and Country in America,* ed. David R. Weimer (New York: Appleton-Century-Crofts, 1962), 310.

81. Fishman, *Urban Utopias,* 163.

82. Ibid., 231.

83. This section draws from a special exhibit at the National Building Museum: *Designing Tomorrow: America's Worlds Fairs of the 1930s.* See www.nbm.org/exhibitions-collections/exhibitions/worlds-fairs.html.

84. Chicago Historical Society, "A Century of Progress, 1933–34," www.chicagohs.org/history/century/cent5.html.

85. As documented in James Mauro, *Twilight at the World of Tomorrow: Genius, Madness, Murder, and the 1939 World's Fair on the Brink of War* (New York: Ballantine Books, 2010), 58.

86. Bennett Johnson, ed., *A Century of Progress 1933–34* (Chicago: Chicago Art Deco Society, 2004), 1.

87. Ibid., 11.

88. Mauro, *Twilight at the World of Tomorrow.*

89. The 1939 site was reused for the 1964 World's Fair, and has accommodated many quasi-public facilities including the Billy Rose Aquacade, the Queens Museum, Shea Stadium (until recently the home of the New York Mets), the Queens Botanical Garden, and the campus of Queens College, along with some actual park areas and several highways.

90. Robert W. Rydell, introduction to *Designing Tomorrow: America's World's Fairs of the 1930s,* ed. Robert W. Rydell and Laura Bird Schiavo (New Haven: Yale University Press, 2010), 1.

91. Lisa D. Schrenk, "'Industry Applies'—Corporate Marketing at A Century of Progress," in *Designing Tomorrow,* ed. Rydell and Schiavo, 23–39.

92. Mauro, *Twilight at the World of Tomorrow,* 143–44.

93. Robert Bennett, "Pop Goes the Future," in Rydell and Schiavo, *Designing Tomorrow,* 177–91, n. 19, 180.

94. E. B. White, "The World of Tomorrow" (1939), reprinted in E. B. White, *One Man's Meat* (New York: Harper & Row, 1978), 66–67.

95. Cited in Mauro, *Twilight at the World of Tomorrow,* 177.

96. In reaction to the patronizing atmosphere of the resulting pavilion, protests from the African-American community persuaded the powers that be to include four murals by the black artist Aaron Douglas.

97. Robert W. Rydell, "Introduction," 12–14.

98. Frank Hobbs and Nicole Stoops, *Demographic Trends of the 20th Century* (Washington, DC: U.S. Census Bureau, 2002), fig. 1.20, www.census.gov/prod/2002pubs/censr-4.pdf.

Part II: The Technocrat Decades, 1945–1990

1. Alan Altshuler and David Luberoff, *Mega-Projects: The Changing Politics of Urban Public Investment* (Washington, DC: Brookings Institution Press, 2003), 13.

2. Adam Rome, *The Bulldozer in the Countryside: Suburban Sprawl and the Rise of American Environmentalism* (New York: Cambridge University Press, 2001), 18.

3. Ibid., 34–35.

4. Rosalyn Baxandall and Elizabeth Ewen, *Picture Windows: How the Suburbs Happened* (New York: Basic Books, 2000), 87–116, as cited in Dolores Hayden, *Building Suburbia* (New York: Vintage Books, 2002), 130.

5. Jon C. Teaford, *The Twentieth-Century American City: Problem, Promise, and Reality* (Baltimore: Johns Hopkins, 1993), 100.

6. Gunnar Myrdal, *An American Dilemma: The Negro Problem and American Democracy* (New York: Harper and Row, 1944), 305. See also Norman Krumholz, "The Kerner Commission Twenty Years Later," in *The Metropolis in Black and White: Place, Power, and Polarization,* ed. George C. Galster and Edward W. Hill (Rutgers Center for Urban Policy Research, 1994), 19–38, 31.

7. Jon C. Teaford, *City and Suburb: The Political Fragmentation of Metropolitan America 1850–1970* (Baltimore: Johns Hopkins University Press, 1979), 5.

3. The Central City Renewal Engine

1. Joseph P. Fried, *Housing Crisis U.S.A.* (New York: Praeger, 1971), 87.

2. Ibid., 66.

3. Quoted ibid., 71–72.

4. FDR quoted in Eugenie L. Birch, "Woman-Made America: The Case of Early Public Housing Policy," in *The American Planner: Biographies and Recollections,* ed. Donald A. Krueckeberg (New York: Methuen, 1983), 170.

5. Ibid.

6. For a perceptive discussion of Moses's public housing projects in New York City see Nicholas Dagen Bloom, *Public Housing That Worked: New York in the Twentieth Century* (Philadelphia: University of Pennsylvania Press, 2008).

7. Ibid, 119.

8. Fried, *Housing Crisis U.S.A.,* 72.

9. Derek S. Hyra, *The New Urban Renewal: The Economic Transformation of Harlem and Bronzeville* (Chicago: University of Chicago Press, 2008). In 2012, Hurricane Sandy devastated many of the Moses-era high-rise public projects with loss of power (and thus elevators) and water.

10. Alexander Polikoff, *Waiting for Gautreaux: A Story of Segregation, Housing, and the Black Ghetto* (Evanston, IL: Northwestern University Press, 2006), 39.

11. Paul S. Grogan and Tony Proscio, *Comeback Cities: A Blueprint for Neighborhood Revival* (Boulder, CO: Westview Press, 2000), 187.

12. According to the description on the Interstate Realty Management Company website: "Central to this ambitious and comprehensive plan [of Legends South] is the integration of

the development site with the surrounding community. The area's original city grid will be re-established and a wide variety of building types, none taller than four stories, will blend seamlessly with the neighborhood. The transformation of State Street into a boulevard and the development of new retail space along neighboring east-west corridors, in addition to the vital component of homeownership, sets the stage for long term private investments, physical and social improvements and the rebirth of one of Chicago's most culturally rich districts" (www.legendssouth.com/).

13. Catherine Bauer, "The Dreary Deadlock of Public Housing," *Architectural Forum*, May 1957, 140.

14. Polikoff served as pro bono attorney directing the litigation from 1966 until 1969, after which he became executive director of the Chicago public interest law firm Business and Professional People for the Public Interest (BPI), where he has waged the fight against discrimination in public housing ever since.

15. See BPI, "Public Housing—*Gautreaux* Litigation," www.bpichicago.org/ph_gautreaux .php.

16. Polikoff, *Waiting for Gautreaux*, 58.

17. *Gautreaux v. Chicago Housing Authority*, 296 F. Supp. 907 (1969) and 304 F. Supp. 736 (1969). As drafted by lawyers for the plaintiffs, Judge Austin's order established a "limited public housing area" consisting of census tracts exceeding 30 percent in black population, plus a one-mile buffer zone beyond such tracts. New units required to be built in "white" neighborhoods were required to be outside those parts of the city.

18. Polikoff, *Waiting for Gautreaux*, 102. Congress in 1974 had established the Section 8 Housing Choice Program, which theoretically enabled low-income households to apply federal rent subsidies to available units anywhere in a metropolitan housing market. See my further discussion in chapter 7.

19. 363 F. Supp. 690 (1973).

20. *Milliken v. Bradley*, 418 U.S. 717 (1974).

21. *Hills v. Gautreaux*, 425 U.S. 284, 1976. According to the front-page headline in the *New York Times*, April 4, 1976, the decision "Approve[d] Housing Poor in White Suburbs," a conclusion Polikoff felt was overstated.

22. Polikoff, *Waiting for Gautreaux*, 150.

23. Ibid., 177.

24. Richard F. Babcock and Charles. L. Siemon, chap. 9, "*Gautreaux*: Chicago's Tragedy," *The Zoning Game Revisited* (Boston: Oelgeschlager, Gunn and Hain, 1985).

25. Polikoff, *Waiting for Gautreaux*, chap. 6.

26. Martha Biondi, "Robert Moses, Race, and the Limits of an Activist State," in *Robert Moses and the Modern City: The Transformation of New York*, ed. Hilary Ballon and Kenneth T. Jackson (New York: W. W. Norton, 2007), 118.

27. Quoted ibid., 117.

28. Ibid., 119.

29. Charles V. Bagli, "Megadeal: Inside a Real Estate Coup," *New York Times*, December 31, 2006.

30. Charles V. Bagli and Christine Haughney, "Fallout Is Wide in Failed Deal for Stuyvesant," *New York Times*, January 26, 2010.

31. See www.pcvstliving.com and www.stpcvta.org.

32. William L. Slayton, "The Operation and Achievements of the Urban Renewal Program" in *Urban Renewal: The Record and the Controversy*, ed. James Q. Wilson (Cambridge: MIT Press, 1971), 192 (emphasis added).

33. Charles Abrams, "Some Blessings of Urban Renewal," in Wilson, *Urban Renewal,* 569.

34. Jon C. Teaford, *The Rough Road to Renaissance: Urban Revitalization in America, 1940–1985* (Baltimore: Johns Hopkins University Press, 1990), 108–9.

35. The Prudential Center was not literally a product of the federal urban renewal program, but it was an exemplar of public-private collaboration to redevelop a downtown site for private purposes using public eminent domain power and tax subsidies. The Massachusetts court initially blocked the project but later allowed it to proceed with a new legislative declaration that "urban renewal is a public purpose." See Charles M. Haar, "The Social Control of Urban Space," in *Cities and Space: The Future Use of Urban Land,* ed. Lowden Wingo Jr. (Baltimore: Johns Hopkins University Press, 1963), 181.

36. Alexander Garvin, *The American City: What Works, What Doesn't* (New York: McGraw-Hill, 2002), 286.

37. *Berman v. Parker,* 348 U.S. 26, 75 S. Ct. 98 (1954).

38. John R. Wennersten, *The Death and Life of an American River* (Baltimore: Chesapeake Book Co., 2008), 147–48.

39. The landmark school desegregation decision in *Brown v. Board of Education* was issued earlier in the same term as *Berman,* but that remarkable decision could scarcely have been foreseen when the latter case was filed.

40. The relevant clause of the Fifth Amendment reads: "Nor shall private property be taken for public use without compensation." While compensation would be paid to the plaintiff property owners, they claimed that the purpose of the "taking" was not, in fact, a "public use" since the site was designated to be redeveloped for other commercial purposes by private redevelopers with the benefit of a huge federal subsidy (write-down) of the sale price.

41. *Berman v. Parker,* at 102–3.

42. Slayton, "Operation and Achievements of the Urban Renewal Program," 203.

43. Garvin, *American City,* 273.

44. Ada Louise Huxtable, *Will They Ever Finish Bruckner Boulevard?* (Berkeley: University of California Press, 1970), 120.

45. William Alonso, "Cities, Planners, and Urban Renewal," in *Urban Renewal: The Record and the Controversy,* ed. James Q. Wilson (Cambridge: MIT Press, 1971), 452. At the time of writing, the Gateway Arch and riverfront in St. Louis is planned for a major upgrade to enhance access and tourism. See Charlene Prost, "Out of Isolation," *Planning* 78.6 (July 2012): 28–31.

46. Alex Marshall, *How Cities Work: Suburbs, Sprawl, and the Roads Not Taken* (Austin: University of Texas Press, 2000), 52.

47. Astute lobbying earned a special credit for university costs in furtherance of an urban renewal plan toward the city's "nonfederal share." Universities also gained access to the power of eminent domain through their respective city renewal authorities.

48. The "Oak Street Connector," as it was originally named, is a "sunken, mile-long stub of 1960s freeway . . . as much a divider as a connector [which is scheduled to be rebuilt as] a pair of boulevards, . . . bike lanes, and wide sidewalks." John Dillon, "Only Connect," *Yale Alumni Magazine* (March/April, 2013), 26.

49. Garvin, *American City,* 260–62. New Haven's Urban Resources Initiative is summarized in chapter 12 of Garvin's book.

50. As of 2011, Columbia University is planning a new campus to be set within nearby Harlem despite strong community opposition. Harvard, for financial reasons, has abandoned

a similar plan to construct a new campus in the industrial neighborhood of Allston across the Charles River in Boston. And New York University is roiling opposition against its proposed expansion into Greenwich Village in Manhattan.

51.   *Harrison-Halsted Community Group, Inc. v. Housing and Home Finance Agency et al.* 310 F.2d 99 (1962). Appeal to the U.S. Supreme Court was denied. The Hull House complex of thirteen buildings (discussed in chapter 2) was sold to the city in the early 1960s. The organization reincorporated as the Hull House Association, which today operates three social service facilities in lower-income neighborhoods of the city. Hull House itself and a dining hall structure were renovated and now serve as the Hull House Historical Museum under the National Park Service.

52.   Douglas Commission [formally National Commission on Urban Problems], *Building the American City* (New York: Praeger, 1969). The commission was known as the Douglas Commission after its chair, Senator Paul H. Douglas.

53.   Cited in Fried, *Housing Crisis U.S.A.,* 88–89.

54.   Herbert J. Gans, "The Failure of Urban Renewal," in Wilson, *Urban Renewal,* 541.

55.   Herbert J. Gans, *The Urban Villagers: Group and Class in the Life of Italian-Americans* (New York: Free Press, 1962); Garvin, *American City,* 272.

56.   Charles Abrams, *The City Is the Frontier* (New York: Harper & Row, 1965), 117.

57.   Douglas Commission, *Building the American City,* 153.

58.   Jane Jacobs, *The Death and Life of Great American Cities* (New York: Random House, 1961), 6.

59.   Douglas Commission, *Building the American City,* 160.

60.   Fried, *Housing Crisis U.S.A.,* 97.

61.   Pub. L. No. 84–627.

62.   Stephen B. Goddard, *Getting There: The Epic Struggle between Road and Rail in the American Century* (New York: Basic Books, 1994), 184.

63.   Ibid., 186.

64.   Quoted in Ibid., 182.

65.   Ibid., 214.

66.   Information accessed April 29, 2011, from the website of Sustainable South Bronx, www.ssbx.org/.

67.   Robert D. Bullard, "The Anatomy of Transportation Racism," in *Highway Robbery: Transportation Racism and New Routes to Equity,* ed. Robert T. Bullard, Glenn S. Johnson, and Angel O. Torres (Cambridge, MA: South End Press, 2004), 19.

68.   Ibid., 19–20.

69.   Omar Freilla, "Burying Robert Moses's Legacy in New York City," in Bullard, Johnson, and Torres, *Highway Robbery,* 75–76.

70.   Roberta Gratz, *The Battle for Gotham: New York in the Shadow of Robert Moses and Jane Jacobs* (New York: Nation Books, 2010), chap. 4.

71.   Ballard and Jackson, *Robert Moses and the Modern City,* 212–16.

72.   O'Neill quote in Helen Leavitt, *Superhighway-Superhoax* (Garden City, NY: Doubleday, 1970), 53.

73.   W. Edward Orser, *The Gwynns Falls: Baltimore Greenway to the Chesapeake Bay* (Charleston, SC: History Press, 2008), 136.

74.   For a discussion of urban and inner city issues during these administrations, see Teaford, *Rough Road to Renaissance,* chap. 5.

75.   Roger Biles, *The Fate of Cities: Urban America and the Federal Government, 1945–2000* (Lawrence: University Press of Kansas, 2011). President Kennedy's chief, direct contribution

to civil rights was the issuance of Executive Order 11063 on November 10, 1962, which prohibited discrimination in housing provided wholly or partially with federal assistance.

76. Ibid., 112.

77. See National Trust for Historic Preservation, *Preservation Leadership Forum*, "Section 4(f) of the Department of Transportation Act," www.preservationnation.org/information-center/law-and-policy/legal-resources/preservation-law-101/federal-law/transportation-act.html#.UVIqPleSi2c.

78. Douglas Commission, *Building the American City*, 155.

79. Polikoff, *Waiting for Gatreaux*, 63, n. 11.

80. Biles, *Fate of Cities*, 141, n. 90.

81. Ibid., 140.

82. Ibid., chap. 11.

4. The Suburban Sprawl Engine

1. The term "suburb" as used here refers to the portions of metropolitan statistical areas (MSAs) located outside of their core city or cities. Thus defined, "suburbs" include traditional bedroom communities, wealthy gated communities, older industrial and blue-collar enclaves, amorphous splotches of postwar housing, shopping malls, and office parks, as well as some remnant villages and farms.

2. Roger Biles, *The Fate of Cities: Urban America and the Federal Government, 1945–2000* (Lawrence: University Press of Kansas, 2011), 150; Robert D. Bullard, Glenn S. Johnson, and Angela O. Torres, eds., *Sprawl City: Race, Politics, and Planning in Atlanta* (Washington, DC: Island Press, 2000), 112.

3. Anthony Flint, *This Land: The Battle over Sprawl and the Future of America* (Baltimore: John Hopkins University Press, 2006), 154.

4. Anthony Downs, *Opening Up the Suburbs* (New Haven: Yale University Press, 1973), 1.

5. Arthur C. Nelson and Thomas W. Sanchez, "Exurban and Suburban Households: A Departure from Traditional Location Theory," *Journal of Housing Research* 8.2 (1997): 249–76, quotation on 251.

6. Paul S. Grogan and Tony Proscio, *Comeback Cities: A Blueprint for Urban Neighborhood Revival* (Boulder, CO: Westview Press, 2000), 9.

7. Jon C. Teaford, *The Twentieth Century American City*, 2nd ed. (Baltimore: Johns Hopkins, 1993), 100.

8. Robert Fishman, *Bourgeois Utopias: The Rise and Fall of Suburbia* (New York: Basic Books, 1987), 182.

9. David Rusk, *Inside Game, Outside Game: Winning Strategies for Saving Urban America* (Washington, DC: Brookings Institution Press, 1999).

10. Calculated by the author using data from "Table of Major Cities: Population," *New York Times Almanac—2009* (New York: Penguin, 2009), 247.

11. William R. Fulton, et al., "Who Sprawls Most? How Growth Patterns Differ Across the United States," Survey Series Monograph (Washington: The Brookings Institution, 2001), 5.

12. Openlands, *Losing Ground: Land Consumption in the Chicago Region, 1900–1998*. Chicago: Openlands, 1998).

13. Fulton et. al., "Who Sprawls Most?"

14. Jean Gottmann, *Megalopolis: The Urbanized Northeastern Seaboard of the U.S.* (Cambridge: MIT, 1961).

15. Ibid., 23. But the geographer John Rennie Short has re-delineated the U.S. Mega-lopolis, using new criteria, to include the four major metro regions of Boston, New York, Philadelphia, and Washington, DC–Baltimore stretching across 13 metropolitan areas in 12 states and 124 counties (plus the District of Colombia) with a combined population of 48.7 million, as compared with 31.9 million in Gottmann's original study area. See John Rennie Short, *Liquid City: Megalopolis and the Contemporary Northeast* (Washington, DC: Resources for the Future Press, 2007).

16. Gottmann, ibid., 15. Poverty and race were of little concern to Gottmann, as reflected in his laconic observation on p. 66: "Megalopolis attracts large numbers of in-migrants from the poorer sections . . . especially Southern Negroes and Puerto Ricans, who congregate in the old urban areas and often live in slums."

17. The editors of the *New York Times,* in "Pulling Back from the Exurbs" (April 10, 2012), quote the Brookings Institute demographer William Frey on exurban growth that "has really come to a standstill and is maybe being given up for dead at this point." The editorial continues: "The country's outer suburbs grew by only 0.4 percent in the fiscal year ended in July [2011]. It peaked in 2006 above 2 percent."

18. Flint, *This Land,* 34.

19. Dolores Hayden, *Building Suburbia: Green Fields and Urban Growth, 1820–2000* (New York: Vintage Books, 2003), chap. 7.

20. Ibid., 135.

21. Reproduced in Adam Rome, *The Bulldozer in the Countryside: Suburban Sprawl and the Rise of American Environmentalism* (New York: Cambridge University Press, 2001), fig. 5.

22. Kenneth T. Jackson, *Crabgrass Frontier: The Suburbanization of the United States* (New York: Oxford University Press, 1985), 237.

23. Ibid.

24. Edward Glaeser, *Triumph of the City* (New York: Penguin, 2011), 175–76.

25. Ibid., 157.

26. William H. Whyte, *The Organization Man* (New York: Simon and Schuster, 1956; republished by University of Pennsylvania Press, 2002). The sociologist Herbert Gans provided a roughly similar appraisal of Levittown, New Jersey, in *Levittowners: Ways of Life and Politics in a New Suburban Community* (New York: Pantheon Books, 1967; rev. ed., 1982).

27. Whyte, ibid., 283–84.

28. Ibid., 311.

29. *Welcome to the Village of Park Forest,* www.villageofparkforest.com.

30. *Levittown, New York,* http://en.wikipedia.org/wiki/Levittown,_New_York# Demographics.

31. Charles Abrams, *Forbidden Neighbors* (New York: Harper & Bros., 1955), 148–49.

32. Jackson, *Crabgrass Frontier,* 203.

33. Abrams, *Forbidden Neighbors,* 229.

34. Hayden, *Building Suburbia,* 132.

35. *Shelley v. Kraemer* 334 US 1 (1948), 20.

36. Marshall also won the landmark school desegregation case in *Brown v. Board of Education* (1954). He was appointed the first African American justice of the U.S. Supreme Court by President Lyndon Johnson, serving from 1967 until 1992.

37. Abrams, *Forbidden Neighbors,* 224–25.

38. Jackson, *Crabgrass Frontier,* 241.

39. *The New York Times Almanac, 2009,* 303.

40. Frank Hobbs and Nicole Stoops, U.S. Census Bureau, "Census 2000 Special Reports," Series CENSR-4. *Demographic Trends in the 20th Century* (Washington, DC: U.S. Government Printing Office, 2002), 130. As of the time of writing, the homeownership rate has dropped by three percentage points since 1970.

41. Tom Daniels, *When City and Country Collide: Managing Growth in the Metropolitan Fringe* (Washington, DC: Island Press, 1999), box 6.1.

42. William W. Goldsmith, "Resisting the Reality of Race: Land Use, Social Justice, and the Metropolitan Economy." Cited in Dolores Hayden, *A Field Guide to Sprawl* (New York W. W. Norton, 2004), 11.

43. Roger Lowenstein, "Who Needs the Mortgage-Interest Deduction?" *New York Times,* March 5, 2006. Lowenstein traces the mortgage interest deduction to the original 1913 federal income tax law as part of a general deduction for business interest. At that time, most home owners paid cash and did not pay mortgage interest. Since the 1940s, as average home size and price have steadily increased, the impact of federal interest and property tax deductions has accordingly grown as a contributing factor in encouraging sprawl.

44. Hayden, *Building Suburbia,* 162.

45. This is not to say that older central cities have entirely failed to repair and update their infrastructure, for example, in the case of New York's third water tunnel (under construction since the 1970s) or two new bridges spanning the Trinity River in Dallas. Frequently, urban infrastructure is built or expanded to benefit both cities and suburbs through regional authorities like the Massachusetts Water Resources Authority, whose new Boston Harbor sewage treatment plant serves Boston and some 42 suburban communities, or counterpart agencies in other parts of the country like the Metropolitan Water Reclamation District of Greater Chicago and the Metropolitan Water District of Southern California.

46. Jeb Brugmann, *Welcome to the Urban Revolution* (New York: Bloomsbury Press, 2009), 257–58.

47. Robert D. Bullard, Glenn S. Johnson, and Angel O. Torres, eds., *Sprawl City: Race Politics and Planning Atlanta* (Washington, DC: Island Press, 2000), 10–11.

48. Harvey K. Flad and Clyde Griffen, *Main Street to Main Frames: Landscape and Social Change in Poughkeepsie* (Albany: SUNY Press, 2009).

49. Stephen B. Goddard, *Getting There: The Epic Struggle between Road and Rail in the American Century* (New York: Basic Books, 1994), 197.

50. A parallel impact of the interstate highway system documented by Stephen B. Goddard (ibid.) has been to subsidize long-haul trucking and bus industries to the detriment of the nation's private freight and passenger railroads. Intercity passenger rail service was largely assimilated into the National Passenger Rail Corporation ("Amtrak") in 1971, whose current prospects are uncertain.

51. "Tax Base Sharing," www.nyslocalgov.org/pdf/Tax_Base_Sharing.pdf.

52. Joel Garreau, *Edge City: Life on the New Frontier* (New York: Doubleday Anchor, 1991).

53. Hayden, *Building Suburbia,* 154.

54. John D. Kasarda and Greg Lindsay, *Aerotropolis: The Way We'll Live Next* (New York: Foster, Strauss, Giroux, 2011), 45.

55. Ibid., 122.

56. Garreau, *Edge City,* 14.

57. Ibid., 46.

58. Brief of Alfred Bettman in *Village of Euclid v. Ambler Realty Co.,* as reprinted in *City and Regional Planning Papers of Alfred Bettman,* Arthur C. Comey. ed. (Cambridge: Harvard University Press, 1946), 157–93.

59. Richard F. Babcock, *The Zoning Game* (Madison: University of Wisconsin Press, 1966), 154.

60. *National Land and Investment Co. v. Easttown Twp. Board of Adjustment* 215 A.2d 597 (1965), at 397.

61. *Appeal of Girsh* 263 A.2d 395 (1970).

62. Ibid., at 399.

63. Norman Williams Jr. and Thomas Norman, "Exclusionary Land Use Controls: The Case of North-Eastern New Jersey," in *Land Use Controls: Present Problems and Future Reform*, ed. Daniel Listokin, (New Brunswick: Rutgers University Center for Urban Policy Research, 1974).

64. 336 A.2d 713 (1975) (*Mount Laurel* I).

65. Ibid., 723.

66. *Southern Burlington County NAACP v. Township of Mount Laurel* 436 A.2d 390 (1983) (*Mount Laurel II*).

67. Richard F. Babcock and Charles L. Siemon, "Chapter 11: Mount Laurel II: Apres Nous le Deluge," in *The Zoning Game Revisited* (Cambridge: Lincoln Institute of Land Policy, 1985), 207–33, quotation on 215.

68. Alan Mallach, "The Betrayal of Mount Laurel," *Shelterforce Online*, March–April 2004, www.nhi.org/online/issues/134/mtlaurel.html.

69. At press time neither case has been decided.

70. Massachusetts General Laws Annotated, Ch. 40B (1969).

71. The state committee reviews each appeal in terms of whether the local zoning rejection is "consistent with local needs," i.e., if any of the following three criteria applies: (1) At least 10 percent of the city's housing stock already is affordable by "low and moderate income" households, (2) subsidized housing occupies more than 1.5 percent of the community's total land area, or (3) the permit will result in construction within one year of more than ten acres or 0.3 percent of the community's land area.

72. Quoted in Brian Mooney, "Debunking Myths Frequently Linked to Affordable Housing," *Boston Globe*, February 18, 2001. In 2010, a statewide voter referendum to repeal the Anti-Snob Zoning Act was rejected by 58 percent of those voting.

5. Battling the Bulldozer

1. Edwin Way Teale, *Dune Boy: The Early Years of a Naturalist* (Bloomington: Indiana University Press, 1943/1957), 2–3.

2. Ibid., 108–109.

3. Henry C. Cowles, "The Ecological Relations of the Vegetation on the Sand Dunes of Lake Michigan," *Botanical Gazette* 27 (May 1899).

4. Harold M. Mayer, "Politics and Land Use: The Indiana Shoreline of Lake Michigan," *Annals of the Association of American Geographers* (1964), 54(4): 508–523, quotation on 508.

5. J. Ronald Engel, *Sacred Sands: The Struggle for Community in the Indiana Dunes* (Middletown, CT: Wesleyan University Press, 1983), 4.

6. Ibid., 57.

7. The popularity of that park would help to support the rest of the Indiana State Park system for decades.

8. Small tracts of this shoreline were occupied by two cottage communities, Ogden Dunes and Dunes Acres, some of whose residents would be leaders in the Save the Dunes movement.

9. William Peeples, "The Indiana Dunes and Pressure Politics," *Atlantic Monthly,* February 1963, 84–88.

10. Ibid., 87.

11. "Indiana Dunes—History," www.savedunes.org/history/.

12. For further discussion see Rutherford H. Platt, *The Open Space Decision Process.* Research Paper 142 (Chicago: University of Chicago Department of Geography Research Series, 1972), 170.

13. Ibid., 159.

14. This was, after all, the era of the hard-knuckle Technocrat Decades—Robert Moses in 1956 bulldozed part of Central Park to create a parking lot for the Tavern on the Green restaurant to the outrage of park advocates.

15. The "compromise" to approve both the park and industrial harbor is attributed to President John F. Kennedy.

16. Further complicating the future of the park, as with Cape Cod and other national seashores, has been the legal practice of buying existing homes within the park boundaries subject to protection of the owner's right of occupancy for life or a period of years (typically 25). Over time, these inholdings, unless renewed, are expected to expire and become publicly owned.

17. Paul H. Douglas, *In the Fullness of Time: The Memoirs of Paul H. Douglas* (New York: Harcourt Brace Jovanovich, 1972), 543.

18. "The Bethlehem Steel plant at Burns Harbor contained the gamut of environmental hazards found in the steel industry. As an integrated steel plant, there were significant emissions and workplace hazards associated with coke ovens; there were waste disposal problems tied to the finishing end of the production process, as well as the use of blast furnaces to extract iron from the ore for steel making." Robert Gottlieb, *Forcing the Spring: The Transformation of the American Environmental Movement* (Washington, DC: Island Press, 1993), 271.

19. Charlotte Read, "The Battle for the Indiana Dunes," chronological history available at www.savedunes.org/history/.

20. Douglas, *In the Fullness of Time,* 540–41. Douglas famously said in 1964: "Until I was thirty, I wanted to save the world. Between the ages of thirty and sixty, I wanted to save the country. But since I was sixty, I've wanted to save the Dunes." Quoted in Engel, *Sacred Sands,* 6.

21. Richard Whitman et al., "The Indiana Dunes: Applications of Landscape Ecology to Urban Park Management," in *The Ecological City: Preserving and Restoring Urban Biodiversity,* ed. Rutherford H. Platt, Rowan A. Rowntree, and Pamela C. Muick (Amherst: University of Massachusetts Press, 1994), 188.

22. National Park Service, "Paul H. Douglas Center for Environmental Education," www.nps.gov/indu/planyourvisit/deec.htm.

23. "Dunes Learning Center," www.duneslearningcenter.org/.

24. Engel, *Sacred Sands,* xviii–xix.

25. Frederick Law Olmsted Jr. and Theodora Kimball, eds., *Forty Years of Landscape Architecture: Central Park* (Cambridge: MIT Press, 1928/1973), 73.

26. For further discussion see Rutherford H. Platt, "Perspectives on Nature in Cities," in Platt et al., *Ecological City,* 23.

27. Engel, *Sacred Sands,* xix.

28. Quoted in Gottlieb, *Forcing the Spring: The Transformation of the American Environmental Movement* (Washington, DC: Island Press, 1993), 28.

29. Hetch Hetchy is a valley in the Sierra Nevada that was dammed to provide a reliable water source to San Francisco after that city's earthquake and fire in 1906. The dam was supported by progressives such as Gifford Pinchot, secretary of the interior under President Theodore Roosevelt, and was finally approved by President Woodrow Wilson in 1913.

30. Stewart L. Udall, *The Quiet Crisis* (New York: Avon Books, 1963).

31. Lewis Mumford, "The Natural History of Urbanization," in *Man's Role in Changing the Face of the Earth,* ed. William L. Thomas Jr., 382–98 (Chicago: University of Chicago Press, 1956), quotation on 386.

32. F. Fraser Darling and John P. Milton, eds., *Future Environments of North America* (Garden City, NY: The Natural History Press, 1966). That conference and book, organized by the biologist Fairfield Osborn, director of the New York Zoological Society, lacked even a Lewis Mumford to express an urban perspective. Four decades later, the New York Zoological Society, renamed the Wildlife Conservation Society, would sponsor Eric Sanderson's *Mannahatta* research and book documenting the "nature" of New York City, which I discuss in chapter 10.

33. William E. Rees, "Understanding Urban Ecosystems: An Ecological Economics Perspective," in *Understanding Urban Ecosystems: A New Frontier for Science and Education,* ed. Alan R. Berkowitz, Charles H. Nilon, and Karen S. Hollweg, 115–36 (New York: Springer-Verlag, 2003), quotation on 118.

34. Gottlieb, *Forcing the Spring,* 242.

35. For further discussion see Rutherford H. Platt, *Land Use and Society: Geography, Law, and Public Policy* (Washington, DC: Island Press, rev. ed. 2004), 410–11.

36. Jonathan Harr, *A Civil Action* (New York: Vintage Books, 1996).

37. Dan Fagin, *Toms River: A Story of Science and Salvation* (New York: Random House Bantam Books, 2013).

38. Robin W. Winks, *Laurance S. Rockefeller: Catalyst for Change* (Washington, DC: Island Press, 1997), 14.

39. Quoted in Thomas Friedman, "G.(reen)O.P.?" Op-ed column, *New York Times,* June 2, 2012.

40. Editors of *Fortune,* eds. *The Exploding Metropolis* (Garden City, NY: Doubleday Anchor, 1958).

41. Charles E. Little, *Challenge of the Land* (New York: Open Space Institute, 1968).

42. William H. Whyte, *The Last Landscape* (Garden City, NY: Doubleday, 1968).

43. Adam Rome, *The Genius of Earth Day: How a 1970 Teach-In Unexpectedly Made the First Green Generation* (New York: Hill and Wang, 2013), 35.

44. Bernice B. Popelka, *Saving Peacock Prairie: The Grassroots Campaign to Protect a Wild Urban Prairie* (Self-published, ca. 2012), 1.

45. Ibid., 2.

46. For further discussion see Rutherford H. Platt, *Open Land in Urban Illinois: Roles of the Citizen Advocate* (DeKalb: Northern Illinois University Press, 1971), 15.

47. Winks, *Laurance S. Rockefeller,* 2.

48. Pub. L. 85-470 (1958). The commission's purpose according to the act's preamble was "to preserve, develop, and attain accessibility to all American people . . . such quality and quantity of outdoor recreation resources . . . to assure the spiritual, cultural, and physical benefits that such outdoor recreation provides; in order to inventory and evaluate the outdoor recreation resources and opportunities of the Nation, [and] to determine the types and location of such resources and opportunities which will be required by present and future generations."

49. Outdoor Recreation Resources Review Commission (ORRRC), *Outdoor Recreation for America* (Washington, DC: Government Printing Office, 1962).

50. *Heart of Atlanta Motel v. U.S.* 379 U.S. 241 (1964).

51. National Outdoor Recreation Act, Pub. L. 88-29 (1963).

52. Land and Water Conservation Fund Act of 1965, Pub. L. No. 88-578. The act required states to prepare "comprehensive outdoor recreation plans" as a basis for allocating LWCF money among state agencies and local communities. The fund was endowed with proceeds from various taxes and fees, and later by royalties from federal Outer Continental Shelf oil and gas leases. Sixty percent of the fund's outlays were allotted to federal agencies for land acquisition for national parks, forests, wildlife refuges, and other facilities. The other 40 percent was allocated by formula to the fifty states to be further divided among state and local agencies for planning, acquisition, and development of sites for outdoor recreation.

53. Winks, *Laurance S. Rockefeller,* 125.

54. Ian McHarg, *Design with Nature* (Garden City, NY: Garden City Press, 1968; repr., New York: Wiley, 1992).

55. Whyte, *The Last Landscape,* 182.

56. Anne Whiston Spirn, *The Granite Garden: Urban Nature and Human Design* (New York: Basic Books, 1984), 5.

6. Legacies of Sprawl

1. Jeb Brugmann, *Welcome to the Urban Revolution: How Cities are Changing the World* (New York: Bloomsbury Press, 2009), xiv–xvi.

2. William H. Whyte, "Urban Sprawl," in Editors of *Fortune*, eds., *The Exploding Metropolis* (Garden City, NY: Doubleday Anchor, 1957).

3. James Howard Kunstler, *The Geography of Nowhere: The Rise and Decline of America's Manmade Landscape* (New York: Simon & Schuster, 1993), 138–39.

4. Robert Bruegmann, *Sprawl: A Compact History* (Chicago: University of Chicago Press, 2005), 11.

5. Ibid., 97.

6. Ibid., 101.

7. Joel Kotkin, *The Next Hundred Million: America in 2050* (New York: Penguin, 2010).

8. For further discussion see Thomas L. Friedman, *Hot, Flat, and Crowded* (New York: Farrar, Straus, and Giroux, 2008).

9. Frank Hobbs and Nicole Stoops, U. S. Census Bureau, Census 2000 Special Reports. Series CENSR-4. *Demographic Trends in the 20th Century* (Washington, DC: U.S. Government Printing Office, 2002), 130, figs. 1-1 and 4-1.

10. Ibid., fig. 5-3.

11. William H. Frey and Alan Berube, "City Families and Suburban Singles: An Emerging Household Story from Census 2000" (Washington, DC: The Brookings Institution, 2002), 1.

12. Sabrina Tavernise, "Married Couples are No Longer a Majority, Census Finds," *New York Times,* May 26, 2011, 8.

13. "News Releases," accessed August 10, 2011, from http://2010.census.gov/news/releases/operations/cb11-cn147.html.

14. Alan Ehrenhalt, *The Great Inversion and the Future of the American City* (New York: Alfred Knopf, 2012), chap. 4.

15. Norimitsu Onishi, "New San Francisco Tech Boom Brings Jobs but also Worries," *New York Times*, June 5, 2012.

16. Ehrenhalt, *The Great Inversion*, 85.

17. Karen Kornblum, "The Parent Trap," *Atlantic Monthly*, January–February 2003, 111.

18. The median price of a new home rose over ten-fold from $23,400 in 1970 to $246,000 by 2006 (before the housing downturn began). See data in chart available at www.census.gov/const/uspriceann.pdf. Meanwhile, the consumer price index for that period rose fivefold.

19. Ibid.

20. Moya Mason, "Housing: Then, Now, and Future," www.moyak.com/papers/house-sizes .html.

21. Kaid Benfield, "What's Going On with new Home Sizes—Is the Madness Over," *Switchboard: Natural Resources Defense Council Staff Blog*, February 9, 2012, http://switchboard .nrdc.org/blogs/kbenfield/us_home_size_preferences_final.html.

22. "Median and Average Square Feet of Floor Area in New Single-Family Houses Completed by Location, www.census.gov/const/C25Ann/sftotalmedavgsqft.pdf.

23. Benfield, "What's Going On."

24. William Fulton, et al., "Who Sprawls Most? How Growth Patterns Differ Across the U.S." The Brookings Institution Survey Series (July 2001), www.brookings.edu/~/media/ research/files/reports/2001/7/metropolitanpolicy%20fulton/fultoncasestudies.pdf.

25. Edward J. Blakely and Mary G. Snyder, *Fortress America: Gated Communities in the United States* (Cambridge, MA: Lincoln of Land Policy and the Brookings Institution, 1997).

26. Quoted in Christopher B. Leinberger, "The Next Slum? McMansions as Tomorrow's Tenements," *Atlantic Monthly*, March 2008, available at www.theatlantic.com.

27. Ibid.

28. Brugmann, *Welcome*, 121–22. Brugmann's assertion is not based solely on North American development failures; in several analyses in his book he draws on examples from around the world. One case that I personally observed in 2007 was a shabby collection of half-finished condos near (but not on) the Mediterranean coast at Antalya, Turkey. They displayed plaintive "For Sale" signs in English and Russian.

29. Bob Herbert, "Losing Our Way," op-ed column, *New York Times*, March 25, 2011.

30. Keith Wardrip, "Housing Landscape" (Center for Housing Policy, 2011), www.nhc .org/publications/index.html. "Working households" are defined by the NHC as those reporting members working at least 20 hours per week, with incomes no higher than 120% of average median income for their area.

31. Ibid., fig. 4.

32. Urban Land Institute, "Priced Out: Persistence of the Workforce Housing Gap in the Boston Metro Area, October 4, 2010, www.chapa.org/news/urban-land-institute-releases- priced-out-report-greater-boston-october-4-2010.

33. Adam Nagourney, "Latest Los Angeles Traffic Plan: Scare off Its Worst Nightmare," *New York Times*, July 7, 2011.

34. Neal Peirce, "The Bottom-Line Reality of Unaffordable Housing" op-ed column, *Seattle Times*, December 6, 2004.

35. *2009 New York Times Almanac*, 421.

36. Like many critical traffic arteries, the New York Thruway Tappan Zee Bridge has been overdue for major repair or replacement for at least a decade. It was built at the widest point on the Hudson River to avoid the clutches of New York City's public works czar, Robert Moses (whose bridges have held up much better than Tappan Zee). See James Panero,

"The Hudson River Destruction Project," *City Journal,* Spring 2011, www.city-journal .org/2011/21_2_hudson-river.html.

37. John Seabrook, "The Slow Lane," *New Yorker,* September 2, 2002, 120–29.

38. Ibid., 123.

39. Robert D. McFadden and Alison Leigh Cowan, "Bumper to Bumper, Travelers Ride out an I-95 Nightmare," *New York Times,* March 27, 2004.

40. Randy Kennedy, "I-95, a River of Commerce Overflowing with Traffic," *New York Times,* December 29, 2000.

41. Ibid.

42. Texas Transportation Institute, Press Release, Jan. 20, 2011.

43. Ibid.

44. Paul Sorenson, "Reducing Traffic Congestion and Improving Travel Options in Los Angeles," www.newgeography.com/content/001318-reducing-traffic-congestion-and-improving-travel-options-los-angeles.

45. Randy Kennedy, "The Day the Traffic Disappeared," *New York Times Magazine,* April 20, 2003, 42–45.

46. Cited in Lester R. Brown, "Designing Cities for People," *Plan B 4.0: Mobilizing to Save Civilization* (New York: W. W. Norton, 2009), 166.

47. "Miasma Theory," http://en.wikipedia.org/wiki/Miasma_theory.

48. In 1854, Dr. John Snow proved that a cholera outbreak in a London neighborhood infected those who used one particular well for water, but spared neighbors who had another water source, thus identifying contaminated waters supplies as the medium of infection. For further discussion see Steven Johnson, *The Ghost Map: They Story of London's Most Terrifying Epidemic and How It Changed Science, Cities, and the Modern World* (New York: Riverhead Books, 2006).

49. Motor Vehicle Air Pollution and public health: Asthma and Other Respiratory Effects (1999), www.edf.org/documents/2655_motorairpollution.asthma.pdf, accessed June 27, 2011.

50. Roni Caryn Rabin, "Asthma Rate Rises Sharply in U.S., Government Says," *New York Times,* May 3, 2011. The CDC report is available at www.cdc.gov/vitalsigns/Asthma/.

51. Natural Resources Defense Council, "New Medical Study Says Diesel Exhaust May Cause Asthma, Not Just Aggravate It," www.nrdc.org/media/pressreleases/020213b.asp.

52. Dennis Creech and Natalie Brown, "Energy Use and the Environment," chapter 8 in *Sprawl City: Race, Politics, and Planning in Atlanta,* ed. Robert D. Bullard, R. D., Glenn S. Johnson, and Angela O. Torres, 187–208 (Washington, DC: Island Press, 2000), quotation on 198.

53. "Sustainable South Bronx," www.ssbx.org/ssbxblog/.

54. Information from www.ccaej.org/environmental-issues/goods-movement.html.

55. Ibid.

56. For further reading see the introduction to Trust for America's Health and Robert Wood Johnson Foundation, "F as in Fat: How Obesity Threatens America's Future" (issue report, 2011), www.healthyamericans.org/report/100/.

57. World Health Organization, "Obesity and Overweight," www.who.int/mediacentre/factsheets/fs311/en/index.html.

58. Ibid., "Personal Commentary" by Dr. David Satcher.

59. Lester R. Brown, *Eco-Economy: Building an Economy for the Earth* (New York: W. W. Norton, 2001), 197.

60. Trust for America's Health, "F is for Fat."

61. Anne C. Lusk, "Promoting Health and Fitness through Urban Design," in *The Humane Metropolis: People and Nature in the 21st Century City,* ed. Rutherford H. Platt, 87–101 (Amherst: University of Massachusetts Press and Lincoln Institute of Land Policy, 2006), quotation on 87.

62. Richard Louv, *Last Child in the Woods: Saving Our Children from Nature-Deficit Disorder* (Chapel Hill, NC: Algonquin Books, 2008), 47.

63. Ibid., 117.

64. Ibid., 35.

65. Peter Harnik and Ben Welle, *From Fitness Zones to the Medical Mile: How Urban Park Systems Can Best Promote Health and Wellness* (Washington, DC: Trust for Public Land, 2011), 5.

66. Ibid.

67. James Vlahos, "Is Sitting a Lethal Activity," *New York Times,* April 14, 2011.

68. Ibid.

69. U.S. Department of Transportation, Bureau of Transportation Statistics (no date) "Journey-to-Work Trends in the United States and Its Major Metropolitan Areas, 1960–1990," http://ntl.bts.gov/DOCS/473.html.

70. Carl Pope, "Solving Sprawl. The Sierra Club Rates the States" (1999), www.sierraclub.org/sprawl/report99/index.asp.

71. For further discussion see Rutherford H. Platt, "Urban Watershed Management: Sustainability One Stream at a Time," *Environment* 48.4 (2006): 26–42.

72. For further discussion see Rutherford H. Platt, "Floods and Man: A Geographer's Agenda," in *Geography, Resources, and Environment,* vol. II, ed. Robert W. Kates and Ian Burton, 28–68 (Chicago: University of Chicago Press, 1986), fig. on 29.

73. Mike Davis, *Ecology of Fear: Los Angeles and the Imagination of Disaster* (New York: Henry Holt, 1998), 59. Since Davis wrote that, the Friends of the Los Angeles River (http://folar.org/) have worked diligently to restore portions of the river to quasi-natural condition as discussed in chapter 9.

74. Project for Public Spaces, "Riverwalk and Waterplace Park," www.pps.org/great_public_spaces/one?public_place_id=86.

75. American Society of Civil Engineers (ASCE) River Restoration Subcommittee on Urban Stream Restoration. *Journal of Hydraulic Engineering,* 129.7 (2003): 491–93.

76. Gretchen C. Daily, ed., *Nature's Services: Societal Dependence on Natural Ecosystems* (Washington, DC: Island Press, 1997); and American Rivers, *Where Rivers are Born: The Scientific Imperative for Defending Small Streams and Wetlands* (Washington, DC: American Rivers, 2003), www.sierraclub.org/watersentinels/downloads/WhereRiversAreBorn.pdf.

77. Anne Whiston Spirn, "Restoring Mill Creek: Landscape Literacy, Environmental Justice, and City Planning and Design," *Landscape Research* 30.3 (July 2005): 395–413.

78. John R. Wennersten. *Anacostia: The Death and Life of an American River* (Baltimore: Chesapeake Book Co., 2008), 162.

79. Thomas E. Dahl and Charles E. Johnson. *Status and Trends of Wetlands in the Conterminous United States, id-1970s to Mid-1980s.* Washington: U.S. Department of the Interior, Fish and Wildlife Service, 1991).

80. American Rivers, *Paving our Way to Water Shortages: How Sprawl Aggravates the Effects of Drought* (Washington, DC: American Rivers and Natural Resources Defense Council, no date).

81. Mike Davis, *Ecology of Fear,* 18.

82. Many kinds of federal disaster assistance are triggered by a Presidential Disaster Declaration. Small grants and other means-based assistance to stricken individuals are

sorely needed. However, the lion's share of federal assistance reimburses eligible states and local governments for much of their costs of disaster response and recovery—in effect reducing their motivation to limit sprawl in hazardous locations. See Rutherford H. Platt, *Disasters and Democracy: The Politics of Extreme Natural Events* (Washington, DC: Island Press, 1999), chap. 3 and conclusion.

83. Ibid., chap. 8.

84. Ibid., chap. 5.

85. For further discussion see Rutherford H. Platt, "The 2020 Water Supply Study for Metropolitan Boston: The Demise of Diversion," *Journal of the American Planning Association* 61.2 (1995): 185–200; Rutherford H. Platt, Paul K. Barten, and Max J. Pfeffer, "A Full, Clean Glass: Managing New York City's Watersheds," *Environment* 42.5 (2000): 8–19.

86. Steven Malanga, "The Muni-Bond Debt Bomb," *City Journal,* Summer 2010.

87. Jon. C. Teaford, *City and Suburb: The Political Fragmentation of Metropolitan America, 1850–1970* (Baltimore: Johns Hopkins, 1979), 6.

88. Information accessed August 4, 2011, from www.weather.com/outlook/weather-news/news/articles/record-heat/.

89. Brown, *Eco-Economy,* 29, fig. 2-1 and fig. 2-2.

90. Andrew Revkin, "Milestone Nears on Curve Charting the Human Imprint on the Atmosphere," http://dotearth.blogs.nytimes.com/2013/05/02/milestone-nears-on-curve-charting-the-human-imprint-on-the-atmosphere/.

91. Justin Gillis, "As Glaciers Melt, Science Seeks Data on Rising Seas," *New York Times,* Nov. 13, 2010.

92. Brown, *Eco-Economy,* 28.

93. Ibid.

94. Felicity Barringer and Kenneth Chang, "Experts See New Normal as a Hotter, Drier West Faces More Huge Fires," *New York Times,* July 1, 2013.

95. Brady Dennis and Meeri Kim, "Western Wildfires' Size, Intensity, and Impact Are Increasing, Experts Say," *Washington Post,* July 7, 2013.

96. See Felicity Barringer, "Homes Keep Rising in the West Despite Growing Wildfire Threat," *New York Times,* July 6, 2013. Nineteen firefighters perished in July 2013 trying to save homes in outlying locations near Prescott, Arizona.

97. Michael Schwirtz, "Sewage Flows after Storm Exposes Flaws in System," *New York Times,* November 20, 2012.

98. Justin Gillis, "Climate Panel Cites Near Certainty on Warming," *New York Times,* August 19, 2013.

Part III: The (More) Humane Decades, 1990–Present

1. Catherine Tumber, *Small, Gritty, and Green: The Promise of America's Smaller Industrial Cities in a Low-Carbon World* (Cambridge: MIT Press, 2012).

2. James Krohe, Jr., "The Incredible Shrinking City," *Planning* 77.9 (Nov. 2011): 10–15.

3. Joel Kotkin, *The New Geography: How the Digital Revolution is Reshaping the American Landscape* (New York: Random House, 2000).

7. Replanting Urbanism in the 1990s

1. "What Is Smart Growth," www.smartgrowthamerica.org/what-is-smart-growth.

2. Two key Supreme Court decisions that rejected specific public restrictions on development of private property were *Nollan v. California Coastal Commission* 107 S. Ct. 3141

(1987) and *Lucas v. South Carolina Coastal Council* 112 S. Ct. 2886 (1992). Although both decisions involved coastal property, *Lucas* in particular blunted the thrust of 1970s-era proposals to broaden the scope of public authority over the use of private property.

3. The conservative tilt of the nation in the 2000 election swept many Smart Growth advocates out of office along with their programs, as in Maryland in 2002 and New Jersey in 2004. Also in 2004, Oregon passed its "Measure 37" limiting the scope of public land use regulation. And in 2011, Florida's new governor Rick Scott "upended the basic tenets of the state's landmark 1985 Growth Management Act" while also gutting the state's Department of Community Affairs and regional planning councils. During the 2012 election, presidential candidate Mitt Romney tried to disown his advocacy of smart growth (and health care) when he was governor of Massachusetts.

4. Donald D. T. Chen, "The Science of Smart Growth," *Scientific American,* Dec. 2000, 84–91, quot. 86.

5. National Association of Home Builders, *NAHB Statement of Policy on Smart Growth* (Washington, DC: NAHB, 1999).

6. Oliver Gillham, *The Limitless City: A Primer on the Urban Sprawl Debate* (Washington, DC: Island Press, 2002), 158.

7. Peter Calthorpe and William Fulton, *The Regional City: Planning for the End of Sprawl* (Washington: Island Press, 2000), 279.

8. Quoted in Chen, 90, note 6.

9. John Rothchild, "A Mouse in the House: Disney Wants to Sell Americans the Ultimate Fantasy: A Utopian Community," *Time,* December 4, 1995, 62.

10. Alex Marshall, *How Cities Work: Suburbs, Sprawl, and the Roads Not Taken* (Austin: University of Texas Press, 2000), 15.

11. Douglas Frantz and Catherine Collins, "It's a Small Town After All," *New York Times,* December 3, 2010.

12. Anthony Flint, *This Land: The Battle over Sprawl and the Future of America* (Baltimore: Johns Hopkins University Press, 2006), 70–74.

13. Rosalie Deane, Executive Director of the Holyoke Housing Authority, interview with author, April 20, 2012.

14. Michael Leccese, "Denver's Stapleton: Green Urban Infill for the Masses?" www.terrain .org/articles/17/leccese.htm.

15. Ken Schroeppel, "#8 Stapleton Redevelopment," http://denverinfill.com/blog/2010/ 01/8-stapleton-redevelopment.html.

16. Calthorpe and Fulton, *The Regional City,* 228.

17. "Welcome to Village Hill Northampton," www.villagehillnorthampton.com/.

18. Alex Marshall observes: "The fatal flaw of New Urbanist design may be that it has killed off the classic suburban backyard. New Urbanists have done this by making a priority out of establishing a ceremonial street facade. They push homes close to the street and ban front driveways and garages. But because the car must still be accommodated, all the support services for it, like driveways and garages, have been pushed into the alley and the back of the house." (Marshall, *How Cities Work,* 30).

19. As of May 2013, there are signs of new investment at Village Hill with an assisted living facility to be built on the site and possible commercial reuse of one of the hospital buildings. The various types of homes have been fully occupied.

20. Joel Kotkin, *The New Geography: How the Digital Revolution is Reshaping the American Landscape* (New York: Random House, 2000), 35.

21. Quoted in John Ingram Gilderbloom, *Invisible City: Poverty, Housing, and New Urbanism* (Austin: University of Texas Press, 2008), 16.

22. Keith Wardrip, "Housing Landscape 2011 (Center for Housing Policy, February 2011)," www.vtaffordablehousing.org/documents/resources/680_Landscape2011brieffinal.pdf.

23. Ibid. "Working households" are defined by the CHP as those reporting members working at least 20 hours per week, with incomes no higher than 120 percent of average median income for their area.

24. After-tax income for the wealthiest one percent of the nation's households between 1979 and 2007 rose by *275%* as compared with 40% for the middle three-fifths of households, and 18% for the poorest one-fifth. Congressional Budget Office, *Trends in the Distribution of Household Income Between 1979 and 2007* (Washington, DC: CBO, Oct. 25, 2011) www.cbo.gov/doc.cfm?index=12485.

25. *Gautreaux v. Chicago Housing Authority* 296 F. Supp. 907 (1969) and 304 F. Supp. 736 (1969).

26. Section 8 vouchers may be either "project based" (limited to particular affordable housing developments) or "tenant based" (usable for any rental unit subject to landlord approval and cost limits).

27. 425 U.S. 284 (1976).

28. Alexander Polikoff, *Waiting for Gautreaux: A Story of Segregation, Housing, and the Black Ghetto* (Evanston, IL: Northwestern University Press, 2006), 246–47.

29. It is not known to the author what "VI" refers to.

30. Gilderbloom, *Invisible City,* 116–17.

31. "HOPE VI," http://en.wikipedia.org/wiki/HOPE_VI#cite_note-HUD2010Budget-7.

32. Karen Hawkons, "Cabrini-Green Housing Complex Finally Closed," www.huffingtonpost.com/2010/12/01/cabrinigreen-housing-comp_n_790292.html. While most large cities embraced HOPE VI, New York City chose instead to renovate and broaden the occupancy of high-rise Moses-era apartment buildings. Under Mayor Michael Bloomberg, the city has "launched a multi-billion dollar 165,000-unit program targeted at both lower- and middle-income New Yorkers. This is consistent with the intent of HOPE VI to deregulate public housing and promote a more entrepreneurial, market-driven culture in public housing management.

33. "The Crisis in Public Housing," editorial, *New York Times,* November 5, 2010.

34. Paul S. Grogan and Tony Proscio, *Comeback Cities: A Blueprint for Urban Neighborhood Revival* (Boulder, CO: Westview Press, 2000), 3–6.

35. The past dichotomy between public and subsidized affordable housing has blurred since the 1990s as many subsidized developments include units available to poor households eligible for Section 8 vouchers. Indeed, as discussed above, HOPE VI has attempted to relocate many public housing tenants in subsidized or other private market housing.

36. For further information see www.dsni.org/.

37. The "Fair Housing Act" was Title VIII of the 1968 Civil Rights Act, Pub. L. 90-284.

38. Pub. L. 90-448.

39. Pub. L. 95-128, Title VIII of the Housing and Community Development Act of 1977.

40. Robert E. Rubin and Michael Rubinger, "Don't Let Banks Turn Their Backs on the Poor," *New York Times,* December 4, 2004.

41. "Why Use It," www.policylink.org/site/c.lkIXLbMNJrE/b.5136941/k.29F6/Why_Use_it.htm.

42. Grogan and Proscio, *Comeback Cities,* 94.

43. "Low Income Housing Credit," http://en.wikipedia.org/wiki/Low-Income_Housing_ Tax_Credit. The New Urbanist community Village Hill in Northampton, MA (discussed earlier in this chapter) was substantially financed with LIHTCs.

44. According to HUD guidelines: "Participating jurisdictions may use HOME funds for a variety of housing activities, according to local housing needs. Eligible uses of funds include tenant-based rental assistance; housing rehabilitation; assistance to homebuyers; and new construction of housing. HOME funding may also be used for site acquisition, site improvements, demolition, relocation, and other necessary and reasonable activities related to the development of non-luxury housing. Funds may not be used for public housing development, public housing operating costs, or for Section 8 tenant-based assistance." U.S. Department of Housing and Urban Development, "Programs of HUD: Major Mortgage, Grant, Assistance, and Regulatory Programs" (2006), 26, http://archives.hud.gov/ pubs/ProgOfHUD06.pdf.

45. Debbie Cenziper and Jonathon Mummolo, "A Trail of Stalled or Abandoned HUD Projects," *Washington Post,* May 15 and 16, 2011, posted at www.washingtonpost.com/ investigations/ (accessed Dec. 18, 2011). The day after the series appeared, a group of bipartisan federal lawmakers called for a Congressional review of the program.

46. Jeb Brugmann, *Welcome to the Urban Revolution* (New York: Bloomsbury Press, 2009), 266, citing Civic Federation, *Tax Increment Financing* (Chicago: Civic Federation, 2007).

47. Atlanta BeltLine, "Funding," http://beltline.org/about/the-atlanta-beltline-project/ funding/.

48. Matthew Charles Cardinale, "BeltLine TADAC Report," www.atlantaprogressivenews .com/interspire/news/2012/08/29/beltline-tadac-report-refine-affordable-housing- program.html.

49. Richard Dye and David Merriman, "Tax Increment Financing: A Took for Local Economic Development," *Land Lines* (magazine of the Lincoln Institute of Land Policy) 18:1 (January, 2006). This study examined data for 247 TIF districts in 100 municipalities within the six-county Chicago metropolitan area.

50. For further information see www.lisc.org/.

51. For further information see www.enterprisecommunity.com.

52. "LISC 2010 Annual Report," www.lisc.org/annualreport/2010/.

53. Grogan and Proscio, *Comeback Cities,* 70.

54. Joel Bookman, personal communication with author, July 12, 2011.

55. This information was obtained during a site visit and meeting with Susan Yanun, Director of LISC Logan Square New Communities Program, December 9, 2011 and www.zapataapartments.org, accessed December 18, 2011.

56. Pub. L. 101-336, 1990.

57. Ibid., sec. 3(1)(A).

58. Ibid., sec. 2(a)(3)

59. "New York and San Francisco Receive World Green Building Council's Government Leadership Awards at UN Climate Change Conference," press release, U.S. Green Building Council, Dec. 5, 2011, www.usgbc.org/.

60. "LEED," www.usgbc.org/DisplayPage.aspx?CMSPageID=1988.

61. Ibid., adapted by the author.

62. Beth Hanson and Sarah Schmidt, eds. *Green Roofs and Rooftop Gardens* (Brooklyn, NY: Brooklyn Botanic Garden, 2012).

63. USGBC, "LEED for Neighborhood Development," www.usgbc.org/DisplayPage .aspx?CMSPageID=148.

64. Benton MacKaye, "An Appalachian Trail: A Project in Regional Planning," *Journal of the American Institute of Architects* 9 (1921): 325–30.

65. Benton MacKaye, *The New Exploration: A Philosophy of Regional Planning* (New York: Harcourt, Brace, 1928).

66. Pub. L. 90-543 (1968).

67. Sandra L. Johnson, "Federal Programs and Legislation," www.americantrails.org/ resources/feds/NatTrSysOverview.html.

68. The Watts letter in the *Chicago Tribune,* September 25, 1963, is credited by Peter Harnik as the birth of the Rail Trail Movement. See Peter Harnik, *Urban Green: Innovative Parks for Resurgent Cities* (Washington, DC: Island Press, 2010), 98.

69. Rutherford H. Platt, *Open Land in Urban Illinois* (DeKalb: Northern Illinois University Press, 1971), 32–37.

70. Harnik, *Urban Green,* 98.

71. Charles E. Little, *Greenways for America* (Baltimore: Johns Hopkins University Press, 1995), 129.

72. For more information see www.nyc.gov/html/dcp/html/mwg/maps_2_1.shtml.

73. David S. Sampson, "Hudson River Greenway: A Regional Success Story" in *Land Use in America,* ed. Henry L. Diamond and Patrick F. Noonan, 50–55 (Washington, DC: Island Press, 1996). For map of the present Greenway, see www.hudsongreenway.ny.gov/ trailsandscenicbyways/watertrail/WTMap.aspx.

74. The Connecticut River Greenway State Park largely resulted from the work of Terry Blunt, a conservationist and land acquisition specialist for the state, and a close friend of mine who sadly and unexpectedly died in 2011.

75. E. Gregory McPherson, "Cooling Urban Heat Islands with Sustainable Landscapes" in *The Ecological City: Preserving and Restoring Urban Biodiversity,* ed. Rutherford H. Platt, Rowan A. Rowntree, and Pamela C. Muick, 151–71 (Amherst: University of Massachusetts Press, 1994).

76. For more information see www.treepeople.org/. North East Trees, Inc. is another LA-based catalyst to tree planting, parks restoration, and watershed revitalization, see www.northeasttrees.org/.

77. Roger S. Ulrich, "Natural versus Urban Scenes: Some Psychological Effects," *Environmental and Behavior* 13 (1981): 523–56.

78. John F. Dwyer, Harbert W. Schroeder, and Paul H. Gobster, "The Deep Significance of Urban Trees and Forests," in Platt et al., *The Ecological City,* 137–50 (Amherst: University of Massachusetts Press, 1994).

79. Sustainable Communities Online, "American Forests' CITYgreen," www.sustainable .org/economy/forestry-a-wood-products/366-american-forests-citygreen.

80. For further information see http://caseytrees.org/about/mission/.

81. Information at www.gatewaygreen.org/resources/ accessed September 20, 2012.

82. Pub. L. 92-500, codified at 33 USCA, Secs. 1251 et seq. The 1972 law was passed by Congress over President Nixon's veto.

83. Pub. L. 96-510, codified at 42 USCA, Secs. 9601, et seq.

84. EPA, "What Is Nonpoint Source Pollution?, http://water.epa.gov/polwaste/nps/whatis .cfm. Efforts by EPA to control nonpoint sources have involved the development of "Total Maximum Daily Load" criteria (TMDLs) for selected water bodies like Chesapeake Bay. The jury is still out on the effectiveness of this approach.

85. For further discussion see Rutherford H. Platt, *Land Use and Society: Geography, Law, and Public Policy*, rev. ed. (Washington, DC: Island Press, 2004), 411.

86. Anthony DePalma, "Superfund Cleanup Stirs Troubled Water," *New York Times,* August 13, 2012.

87. Kim Martineau, "Newtown Creek Clean-UP Polluting the Air," http://blogs.ei .columbia.edu/2012/09/10/airborne-bacteria-above-newtown-creek-superfund-site-linked-to-pollution-in-water-researchers-say.

88. This standard definition was developed jointly by the U.S. Environmental Protection Agency and the U.S. Army Corps of Engineers.

89. John and Mildred Teal, *Life and Death of the Salt Marsh* (New York: Ballantine Books, 1969).

90. *Morris County Land Improvement Co. v. Parsippany-Troy Hills Twp.* 40 N.J. 539, 192 A.2d 232 (1963).

91. *Turnpike Realty Co. v. Town of Dedham* 362 Mass. 221, 284 N.E.2d 891 (1972).

92. *Just v. Marinette County* 56 Wis. 2d. 7, 201 N.W.2d 761 (1972).

93. 33 USCA Sec. 1344.

94. *NRDC v. Callahan* [Secretary of the Army] 392 F. Supp. 685 (1975).

95. 33 *Code of Federal Regulations* 323.2.

96. Rutherford H. Platt, "Coastal Wetland Management: The Advance Designation Approach," *Environment* 29.9 (November 1987): 16–20; 38–43.

97. U. S. Congress, *A Unified National Program for Managing Flood Losses.* House Document 465 (89th Cong., 2d Sess.) Washington, DC: U.S. Government Printing Office, 1966.

98. Private insurance companies have long shunned offering coverage against flood losses at affordable rates (due to the likelihood of losses). The NFIP was thus designed to offer coverage against damage to structures and their contents, usually at "actuarial" rates paid by the insured property owner.

99. Eric Lipton, "Flood Insurance, Already Fragile Faces New Stress," *New York Times,* November 13, 2012.

100. 16 USCA Secs. 1631–1643.

101. "A Tough Environmental Law May Get Tougher," *Boston Globe,* January 4, 1993, 1.

102. The National Marine Fisheries Service administers the ESA for aquatic species.

103. Endangered Species Act, 16 USCA, Sec, 1632(b).

104. Thomas D. Feldman and Andrew E. G. Jonas, "Sage Scrub Revolution? Property Rights, Political Fragmentation, and Conservation Planning under the Federal Endangered Species Act," *Annals of the Association of American Geographers* 90.2 (2000): 256–92.

105. Reid Ewing, Rolf Pendall, and Donald Chen, *Measuring Sprawl and Its Impacts* (Washington, DC: Smart Growth America, 2003), 4.

106. Endangered Habitats League, *EHL Newsletter,* Fall 2003, 1.

8. New Age "Central Parks"

1. William Muir Whitehill, *Boston: A Topographical History,* 2nd ed. (Cambridge: Harvard University Press, 1968), 35.

2. Frederick Law Olmsted Jr. and Theodora Kimball, eds., *Forty Years of Landscape Architecture: Central Park* (Cambridge: MIT Press, 1928/1973), 73.

3. Elizabeth Stevenson, *Park Maker: A Life of Frederick Law Olmsted* (New York, MacMillan, 1977), 298.

4. Ibid., 309.

5. After Frederick Law Olmsted Sr. withdrew from active park design in the 1890s, his firm under the direction of his son (F.L.O. Jr.) played a central role in such landmark park plans as the 1902 McMillan Commission parks plan for Washington, DC, and the 1909 *Plan of Chicago*.

6. Peter Harnik, "Review of *Millennium Park: Creating a Chicago Landmark* by Timothy J. Gilfoyle," *Journal of Urban Affairs* (August 2008), 30(3): 351–53.

7. Lois Wille, *Forever Open, Clear, and Free: The Struggle for Chicago's Lakefront* (Chicago: University of Chicago Press, 1972), chap. 4. The railroad built its downtown terminal on landfill that it constructed in the 1850s with city and state approval.

8. Ward acceded to the location of the Art Institute within the "forever open" easement but blocked proposals for an armory and other buildings on the restricted land (ibid., 75). The Field Museum of Natural History was located just south of the "forever open" easement area as were other museums and the Soldiers Field stadium (a neoclassic structure now topped by a "space ship" of expensive sky boxes). Under Mayor Richard M. Daley, South Shore Drive was relocated to enhance the redesign of the cluster of buildings south of Grant Park as a "Museum Campus."

9. Janice Metzger, *What Would Jane Say? City-Building Women and a Tale of Two Chicago's* (Chicago: Lake Claremont Press, 2010), 124.

10. Ibid., 126.

11. Administratively, Millennium Park is a public-nonprofit joint venture. The city's Department of Cultural Affairs owns the park and oversees its program of special events, concerts, and festivals. The park is administered by a nonprofit corporation, Millennium Park, Inc., which in turn employs consultants and contractors to maintain and protect the park and its various elements. See http://millenniumpark.net/. The Chicago Parks District is not directly involved in Millennium Park, but in 2012 it began construction of Maggie Daley Park, which will adjoin and complement Millennium Park with facilities for more active outdoor recreation. See: http://maggiedaleyparkconstruction.org/.

12. Harnik, "Review of *Millennium Park*."

13. Peter Harnik, *Urban Green: Innovative Parks for Resurgent Cities* (Washington, DC: Island Press, 2010), 57.

14. Harry Wiland and Dale Bell, *Edens Lost and Found: How Ordinary Citizens Are Restoring Our Great American Cities* (White River Junction, VT: Chelsea Green, 2006), 32–38, quote on 32.

15. The parking garage was later sold by the city to a private corporation (Morgan Stanley) for a one-time payment with no revenue for operation of the park.

16. Edward Uhler, Chief Executive Officer, Millennium Park, Inc., Personal communication, May 8, 2013.

17. Cheryl Ken, *Millennium Park Chicago* (Evanston, IL: Northwestern University Press, 2011), 38.

18. Ibid.

19. Ibid.

20. Wille, *Forever Open*.

21. Ann L. Buttenwieser, *Manhattan Water-Bound*, 2nd ed. (Syracuse: Syracuse University Press, 1999), 28–29.

22. Phillip Lopate, *Waterfront: A Walk Around Manhattan* (New York: Anchor Books, 2004), 30.

23. Ibid., 37.

24. Robert F. Wagner Jr., "New York City Waterfront: Changing Land Use and Prospects for Redevelopment," in *Urban Waterfront Lands* (Washington, DC: National Academy Press, 1980), 84.

25. Lopate, *Waterfront,* 93.

26. *Sierra Club v. U.S. Army Corps of Engineers* 546 F. Supp. 1225 (1982), at 1229.

27. Sam Roberts, "After 20 Years of Delays a River park Takes Shape," *New York Times,* May 16, 2006.

28. Buttenwieser, *Manhattan Water-Bound,* 209.

29. Information from www.hudsonriverpark.org/estuary/index.asp, accessed January 16, 2012. Unfortunately, the Hudson River Park has encountered some severe challenges. City and state funding for the park shrank from $42 million in 2008 to $7 million in 2011. Concurrently, the HRP Trust that administers the park is being sued by Chelsea Piers, a private recreational complex, to repair damage to three piers leased from the Trust caused by marine-boring organisms. The HRP Trust also is faced with heavy costs to repair other deteriorating piers, including the 14-acre Pier 40, the park's primary source of parking revenue, whose steel pilings need replacement at an estimated cost of $90 million. The HRP has already cost $340 million and another $200 million is needed to complete it. Charles V. Bagli and Lise W. Foderara. "Times and Tides Weigh on Hudson River Park," *New York Times,* January 28, 2012, A19. HRP experienced severe flooding in Hurricane Sandy in October, 2012 which destroyed its electrical generators and left the park in darkness and its indoor facilities dependent on emergency power for several month.

30. The High Line concept was modeled on the Promenade Plantée in Paris.

31. Friends of the High Line, "High Line," www.thehighline.org/.

32. Engineering work to execute the design ironically required removal of the wild plants that inspired the project; seeds from existing vegetation were preserved to replant the High Line with some of the same plant species. According to the *New York Times,* the challenge in creating the High Line has been to achieve "a balance between preserving what one called 'the romance of the ruin'—wild grasses growing up through the metal skeleton of rails and rivets—and creating a fresh green corridor pedestrians." Robin Pogrebin, "Designer Detail an Urban Oasis 30 Feet Up," *New York Times,* April 19, 2005, 20.

33. "Year in Photos 2011," www.thehighline.org/blog/2011/12/31/2011-the-year-in-photos.

34. Mass. Department of Transportation and Artery Business Committee, "Looking at the Central Artery/Tunnel Project," undated brochure circulated by the *Boston Globe* and other area newspapers.

35. Sean P. Murphy, "Big Dig's Red Ink Engulfs State," *Boston Globe,* July 17, 2008.

36. Noah Bierman, "Not-So-Green-Acres," *Boston Globe,* July 13, 2008.

37. Ibid.

38. Casey Ross, "Greenway Planners Shifting Approach," *Boston Globe,* May 16, 2010.

39. "The Rose F. Kennedy Greenway," www.rosekennedygreenway.org/.

9. Reclaiming Urban Waterways

1. Anne Whiston Spirn, "Restoring Mill Creek: Landscape Literacy, Environmental Justice, and City Planning and Design," *Landscape Research* 30.3 (July 2005): 395–413.

2. See www.nps.gov/ncrc/programs/rtca/index.htm.

3. See http://water.epa.gov/type/watersheds/index.cfm.

4. See http://water.usgs.gov/wsc/software.html.

5. See www.cwp.org/.

6. Ann L. Riley, *Restoring Streams in Cities: A Guide for Planners, Policy Makers, and Citizens* (Washington, DC: Island Press, 1998).

7. Rutherford H. Platt, "Urban Watershed Management: Sustainability, One Stream at a Time," *Environment* 48.4: 27–42, 33–34.

8. "Natural valley storage," a term devised by Charles River advocates, refers to the capacity of freshwater wetlands to retain storm runoff like a sponge, reducing the rate and volume of downstream flow and, in some locations, replenishing local groundwater through seepage into aquifers.

9. One of the nation's leading state court decisions upholding local floodplain regulation involved one of these sites. Ironically, after the Massachusetts Supreme Judicial Court in *Turnpike Realty Inc. v. Town of Dedham* 184 N.E.2d 891 (1972) held that a town could prohibit building in a wetland without compensating the owner, the site was nevertheless purchased under the Charles River Natural Valley Storage Program.

10. Most Massachusetts towns and cities have established local conservation commissions, which, among other functions, administer the state Wetlands Protection Act (MGLA Ch. 131, Ch. 40A). The act requires any proposal to alter wetlands (defined broadly) to apply to the local conservation commission, which issues an "order of conditions" pursuant to a public hearing, regulating how the proposed project should be conducted to minimize harm to wetland resources and services.

11. Robert Zimmerman, Executive Director of CRWA, personal communication with author, February 1, 2006.

12. Robert Glennon, *Water Follies* (Washington, DC: Island Press, 2002), 109–11.

13. Uwe Steven Brandes, "Bankside Washington, DC," in *Rivertown: Rethinking Urban Rivers,* ed. Paul Stanton Kibel, 47–65, (Cambridge: MIT Press, 2007) quot. 47; Robert Zimmerman, "Put It in Our Backyard: Water, Science and Sustainable Development," *River Voices* (Newsletter of the River Network) 15.3: 36–39.

14. John R. Wennersten. *Anacostia: The Death and Life of an American River* (Baltimore: Chesapeake Book Co., 2008), 250.

15. Rutherford H. Platt et al., "Urban Stream Restoration," chap. 7 in *Growing Greener Cities,* ed. Eugenie L. Birch and Susan M. Wachter (Philadelphia: University of Pennsylvania Press, 2008), 132–33.

16. "Anacostia Watershed Society," www.anacostiaws.org.

17. "Earth Conservation Corps," www.earthconservationcorps.org/#!about/c20r9.

18. Brett Williams, "Life Inside a Watershed: The Renewal of the Anacosta River?," *Newsletter of the Association of American Geographers,* January 2010, 9.

19. Metropolitan Washington Council of Governments, "Anacostia Watershed Restoration Progress and Conditions Report" (May 1998).

20. District of Columbia, "Memorandum of Understanding: Anacostia Watershed Initiative" (March 22, 2000). The *Washington Post* devoted a five-part series to the Anacostia and the AWI in July 2004.

21. Williams, "Life Inside a Watershed," 9.

22. Anacostia Waterfront Initiative (AWI), "The Southwest Waterfront Development Plan and AWAI Vision (District of Columbia Department of Planning, 2003), 1.

23. Summarized from Brandes, "Bankside Washington, DC," 48–49.

24. *Berman v. Parker,* 75 S. Ct. 98 (1954), at 102–3.

25. David Moffat, "Development Plan and AWI Vision for the Southwest Waterfront," places.designobserver.com/media/pdf/Development_Pl_1141.pdf.

26. Wennersten, *Anacostia,* 158.

27. Anacostia Waterfront Initiative (AWI), *The Southwest Waterfront Development Plan and AWI Vision* (February 6, 2003).

28. "Anacostia Waterfront Corporation," http://en.wikipedia.org/wiki/Anacostia_Waterfront_Corporation.

29. Debbi Wilgoren, "Hope, Fret, along the Anacostia," *Washington Post,* July 18, 2004.

30. Anne Olson and Aaron Tuley, "The Future of Houston's Buffalo Bayou," unpublished working paper prepared for the Ecological Cities Project, University of Massachusetts, Amherst (September 2004).

31. Buffalo Bayou Partnership, *Buffalo Bayou and Beyond: Visions, Strategies, Actions for the 21st Century* (Houston: BBP, 2002), 3.

32. Ibid.

33. Olson and Tuley, "The Future of Houston's Buffalo Bayou," 10.

34. "Sabine to Bagby Trail," www.buffalobayou.org/sabinebagby.html.

35. Zain Shauk, "Project to Transform Stretch of Buffalo Bayou into 'Jewel'," www.chron.com/news/houston-texas/article/Project-to-transform-stretch-of-Buffalo-Bayou-2406589.php.

36. "Buffalo Bend Nature Park," www.buffalobayou.org/buffalobendcap.html.

37. Olson and Tuley, "The Future of Houston's Buffalo Bayou."

38. Tad Friend, "River of Angels," *New Yorker,* January 26, 2004, 42–49, quotation on 42.

39. Jennifer Wolch, "Green Urban Worlds," *Annals of the Association of American Geographers* 97.2 (2007): 373–84, quot. 374.

40. Jennifer Price, "Confluence Park, Los Angeles," www.arroyoseco.org/Confluence.htm.

41. Ibid.

42. Joe Linton, *Down by the Los Angeles River* (Berkeley: Wilderness Press, 2005).

43. *Los Angeles River Master Plan,* http://ladpw.org/wmd/watershed/LA/LA_River_Plan.cfm.

44. Lorelei Laird, "Restoring the Water Freeway," *Planning* 78.1 (Jan. 2012): 26–31, quotation on 29.

45. Tony Chavira, "The Los Angeles River Revitalization," *FourStory,* http://fourstory.org/features/story/the-los-angeles-river-revitalization.

46. "Los Angeles River Revitalization Corporation," http://thelariver.com/revitalization/los-angeles-river-revitalization-corporation/.

47. Los Angeles River Revitalization Corporation, "In the Works," http://thelariver.com/revitalization/other-projects/in-the-works/.

48. "Community-Based Park Designs," www.amigosdelosrios.org/community-outreach-2/.

49. "Emerald Necklace," www.amigosdelosrios.org/features/stewards-of-the-emerald-necklace/. The term "Emerald Necklace" relates to the Olmsted Brothers plan for the Los Angeles region in the 1930s, as well as to Olmsted Sr.'s Boston "Emerald Necklace" Plan of the 1880s.

10. Humane Urbanism at Ground Level

1. Eric W. Sanderson, *Mannahatta: A Natural History of New York City* (New York: Abrams, 2009).

2. The book was published to commemorate the four-hundredth anniversary of Hudson's arrival in what would become New York Harbor.

3. Sanderson, *Mannahatta,* 13, 16.

4. Lewis Mumford, "The Natural History of Urbanization," in *Man's Role in Changing the Face of the Earth,* W. L. Thomas, ed., quoted on 386 (Chicago: University of Chicago Press, 1956).

5. Michael Klemens, "Balancing Biodiversity and Land Use Planning," *Westchester Gannett Newspaper,* August 26, 2003.

6. For further discussion see Rutherford H. Platt, "Toward Ecological Cities: Adapting to the 21st Century Metropolis," *Environment* 46.5 (June 2004): 10–27, especially 20.

7. Harvey K. Flad and Clyde Griffen, *Main Street to Mainframes: Landscape and Social Change in Poughkeepsie* (Albany: State University of New York Press, 2009).

8. Ibid., 327–28.

9. This case study is based on Harvey K. Flad and Craig M. Dalton, "A Tree and Its Neighbors: Creating Community Open Space," *Hudson River Valley Review* 21.2 (Spring 2005): 56–67.

10. Rutherford Platt, *American Trees: A Book of Discovery* (New York: Dodd, Mead, 1962), 36.

11. Flad and Dalton, "A Tree and its Neighbors," 57–58.

12. Alexander Garvin, *The American City: What Works, What Doesn't* (New York: McGraw-Hill, 2002).

13. Charlotte Dillon, "1 in 4 Live in Poverty in New Haven," www.yaledailynews.com/news/2010/oct/05/1-in-4-live-in-poverty-in-new-haven/.

14. Yale School of Forestry and Environmental Studies, Urban Resources Initiatives," http://environment.yale.edu/uri/programs/.

15. URI's open access teaching modules are available at http://environment.yale.edu/uri/programs/other-programs/.

16. Catherine Tumber, *Small, Gritty, and Green: The Promise of America's Smaller Industrial Cities in a Low-Carbon World* (Cambridge: MIT Press, 2012), 85.

17. Ibid.

18. Ramiro Devaro-Comas, personal communication, March 6, 2012.

19. Ibid.

20. Adapted from Rutherford H. Platt, "Urban Watershed Management: Sustainability One Stream at a Time," *Environment* 48.4 (May 2006): 26–42, quot. 32.

21. Susan Hines, "Stone Soup," *Landscape Architecture* 95.6 (June 2005): 124–31, quot. 129.

22. Washington Parks and People, "Marvin Gaye Park & the Down by the Riverside Campaign," www.washingtonparks.net/marvin_gaye_park.

23. According to the GP website, the complex today has over 15,000 pots of herbs, salad mix, and other vegetables, thousands of fish, and a livestock inventory of chickens, goats, ducks, rabbits, and bees. See http://growingpower.org/.

24. "Will Allen to Receive NEA Foundation Award for Outstanding Service to Public Education," press release January 23, 2012, www.growingpower.org/blog/.

25. Will Allen, "Make History, Be Part of the First Vertical Farm," www.growingpower.org/blog/make-history-be-part-of-the-first-vertical-farm. For further information, see Will Allen and Charles Wilson, *The Good Food Revolution: Growing Healthy Food, People, and Communities* (New York: Gotham Books, 2012).

26. Michael Houck, personal communication with author, February 22, 2012. The mural was a joint project of Houck's Green Spaces Institute and ArtFX Murals.

27. Michael C. Houck, *Wild in the City: A Guide to Portland's Natural Areas* (Portland: Portland Historical Society, 2000), 68.

28. Ibid., 71.

29. Michael C. Houck and M. J, Cody, eds., *Wild in the City: Exploring the Intertwine* (Corvallis: Oregon State University Press, 2011), 76.

30. Ibid., 72.

31. Michelle Bussard, "Stories from the 'Shed," *Within Your Reach* (Newsletter of the Johnson Creek Watershed Council), Fall 2005.

32. Stephen Johnson, "Co-Production of Public and Civic Actions: A Case Study of Citizen Participation in the Johnson Creek Watershed," unpublished paper presented at International Rivers Symposium, Brisbane, Australia, September 2005.

33. Johnson Creek Watershed Council, "Johnson Creek State of the Watershed," http://jcwc.org/wp-content/uploads/2012/05/2012StateoftheWatershed.pdf.

34. American Museum of Natural History, "Center for Biodiversity and Conservation," www.amnh.org/our-research/center-for-biodiversity-conservation.

35. Chicago Wilderness Home Page, www.chicagowilderness.org/.

36. GreenCityBlueLake Home Page, www.gcbl.org/.

37. Partners for Livable Communities, "Culture Connects All: Rethinking Audiences in Times of Demographic Change" (2011), http://livable.org/livability-resources/reports-a-publications.

38. For further discussion see Rutherford H. Platt, "Botanical Garden Outreach," *Public Garden* 25.1: 5–6.

39. Metropolitan Water Alliance, "About Us," www.waterfrontalliance.org/about.

40. For further discussion see Rutherford H. Platt, "The Humane Megacity: Transforming New York's Waterfront," *Environment* 51.4 (July/August 2009): 46–59, especially 48.

41. Metropolitan Waterfront Alliance, "The Waterfront Action Agenda," www.waterfrontalliance.org/actionagenda.

42. For further discussion see Platt, "The Humane Megacity," 59.

43. See http://chicagowilderness.org/.

44. LEAP, "About LEAP," http://leapbio.org/about.

45. Ibid.

46. Houston Wilderness, "Mission Statement," http://houstonwilderness.org/index.php/about.

47. Houck and Cody, *Wild in the City*.

Epilogue

1. Andrew Light, "Ecological Citizenship: The Democratic Promise of Restoration," in *The Humane Metropolis: People and Nature in the 21st Century*, ed. Rutherford H. Platt, 169–81 (Amherst: University of Massachusetts Press, 2006).

Further Reading

Ballon, Hilary, and Kenneth T. Jackson, eds. *Robert Moses and the Modern City: The Transformation of New York.* New York: Norton, 2007.

Beatley, Timothy. *Green Urbanism: Learning from European Cities.* Washington, DC: Island Press, 2000.

———. *Biophilic Cities.* Washington, DC: Island Press, 2011.

Biles, Roger. *The Fate of the Cities: Urban America and the Federal Government, 1945–2000.* Lawrence: University Press of Kansas, 2011.

Birch, Eugenie L., and Susan M. Wachter, eds. *Growing Greener Cities: Urban Sustainability in the Twenty-First Century.* Philadelphia: University of Pennsylvania Press, 2008.

Bloom, Nicholas Dagen. *Public Housing That Worked: New York in the Twentieth Century.* Philadelphia: University of Pennsylvania Press, 2008.

Brugmann, Jeb. *Welcome to the Urban Revolution.* New York: Bloomsbury Press, 2009.

Bullard, Robert T., Glenn S. Johnson, and Angel O. Torres, eds. *Highway Robbery: Transportation Racism and New Routes to Equity.* Cambridge, MA: South End Press, 2004.

Calthorpe, Peter, and William Fulton. *The Regional City: Planning for the End of Sprawl.* Washington, DC: Island Press, 2000.

Ehrenhalt, Alan. *The Great Inversion and the Future of the American City.* New York: Knopf, 2012.

Engel, J. Ronald. *Sacred Sands: The Struggle for Community in the Indiana Dunes.* Middletown, CT: Wesleyan University Press, 1983.

Flint, Anthony. *This Land: The Battle over Sprawl and the Future of America.* Baltimore: Johns Hopkins University Press, 2006.

Garvin, Alexander. *The American City: What Works and What Doesn't.* New York: McGraw-Hill, 2002.

Gilderbloom, John Ingram. *Invisible City: Poverty, Housing, and New Urbanism.* Austin: University of Texas Press, 2008.

Glaeser, Edward. *Triumph of the City.* New York: Penguin, 2011.

Grogan, Paul S., and Tony Proscio. *Comeback Cities: A Blueprint for Neighborhood Revival.* Boulder, CO: Westview Press, 2000.

Harnik, Peter. *Urban Green: Innovative Parks for Resurgent Cities.* Washington, DC: Island Press, 2010.

Hayden, Dolores. *Building Suburbia: Green Fields and Urban Growth, 1820–2000.* New York: Vintage, 2003.

Jackson, Kenneth T. *Crabgrass Frontier: The Suburbanization of the United States.* New York: Oxford University Press, 1985.

Kibel, Paul Stanton, ed. *Rivertown: Rethinking Urban Rivers.* Cambridge: MIT Press, 2007.

Knight, Louise W. *Jane Addams: Spirit in Action.* New York: Norton, 2010.

Kotkin, Joel. *The New Geography: How the Digital Revolution Is Reshaping the American Landscape.* New York: Random House, 2000.

———. *The Next Hundred Million: America in 2050.* New York: Penguin, 2010.

Little, Charles E. *Greenways for America.* Baltimore: Johns Hopkins University Press, 1995.

Louv, Richard. *Last Child in the Woods: Saving Our Children from Nature-Deficit Disorder.* Chapel Hill, NC: Algonquin Books, 2005.

Marshall, Alex. *How Cities Work: Suburbs, Sprawl, and the Roads Not Taken.* Austin: University of Texas Press, 2000.

Peirce, Neal, and Curtis W. Johnson. *Century of the City: No Time to Lose.* New York: The Rockefeller Foundation, 2008.

Platt, Rutherford H., ed. *The Humane Metropolis: People and Nature in the 21st Century City.* Amherst: University of Massachusetts Press and Lincoln Institute of Land Policy, 2006.

Platt, Rutherford H., Rowan A. Rowntree, and Pamela C. Muick, eds. *The Ecological City: Preserving and Restoring Urban Biodiversity.* Amherst: University of Massachusetts Press, 1994.

Polikoff, Alexander. *Waiting for Gautreaux: A Story of Segregation, Housing, and the Black Ghetto.* Evanston, IL: Northwestern University Press, 2006.

Rome, Adam. *The Bulldozer in the Countryside: Suburban Sprawl and the Rise of American Environmentalism.* New York: Cambridge University Press, 2001.

———. *The Genius of Earth Day: How a 1970 Teach-In Unexpectedly Made the First Green Generation.* New York: Hill and Wang, 2013.

Sanderson, Eric W. *Mannahatta: A Natural History of New York City.* New York: Abrams, 2009.

Short, John Rennie. *Liquid City: Megalopolis and the Contemporary Northeast.* Washington, DC: Resources for the Future Press, 2007.

Spirn, Anne Whiston. *The Granite Garden: Urban Nature and Human Design.* New York: Basic Books, 1984.

Teaford, Jon C. *The Rough Road to Renaissance: Urban Revitalization in America, 1940–1985.* Baltimore: Johns Hopkins University Press, 1990.

———. *The Twentieth-Century American City,* 2nd ed. Baltimore: Johns Hopkins University Press, 1993.

Tumber, Catherine. *Small, Gritty, and Green: The Promise of America's Smaller Industrial Cities in a Low-Carbon World.* Cambridge: MIT Press, 2012.

Wiland, Harry, and Dale Bell. *Edens Lost and Found: How Ordinary Citizens Are Restoring Our Great American Cities.* White River Junction, VT: Chelsea Green, 2006.

Index